Robert Bell

The Geology of Ontario

with special reference to economic minerals

Robert Bell

The Geology of Ontario
with special reference to economic minerals

ISBN/EAN: 9783744717755

Printed in Europe, USA, Canada, Australia, Japan

Cover: Foto ©berggeist007 / pixelio.de

More available books at **www.hansebooks.com**

THE

GEOLOGY OF ONTARIO,

WITH SPECIAL REFERENCE TO

ECONOMIC MINERALS.

BY

ROBERT BELL, B.A.Sc., M.D., LL.D.,

Assistant Director of the Geological Survey of Canada, Member of the Royal Commission on the Mineral Resources of Ontario, formerly Professor of Natural Sciences in Queen's University, etc.

(REPRINTED FROM THE REPORT OF THE ROYAL COMMISSION.)

TORONTO:
PRINTED BY WARWICK & SONS, 68 AND 70 FRONT STREET WEST,
1889.

THE GEOLOGY OF ONTARIO,

WITH SPECIAL REFERENCE TO

ECONOMIC MINERALS.

BY

ROBERT BELL, B.A.Sc., M.D., LL.D.,

Assistant Director of the Geological Survey of Canada, Member of the Royal Commission on the Mineral Resources of Ontario, formerly Professor of Natural Sciences in Queen's University, etc.

(REPRINTED FROM THE REPORT OF THE ROYAL COMMISSION.)

TORONTO:

PRINTED BY WARWICK & SONS, 68 & 70 FRONT STREET WEST.

1889.

THE GEOLOGY OF ONTARIO,

WITH SPECIAL REFERENCE TO ECONOMIC MINERALS.

BY ROBERT BELL, B.A.Sc., M.D., LL.D.,

ASSISTANT DIRECTOR OF THE GEOLOGICAL SURVEY OF CANADA, MEMBER OF THE ROYAL COMMISSION ON THE MINERAL RESOURCES OF ONTARIO, ETC., ETC.

The following sketch of the Geology of Ontario being intended for the use of persons who may not be familiar with the technical terms of geological science, the writer has endeavoured to avoid these as much as possible, but where it has been necessary to employ them their meanings have been briefly given. For the same reason, some elementary geological explanations have been incorporated, and a short glossary of technical words added at the end of the report, to save the non-scientific reader the trouble of referring to geological manuals or text-books. On the other hand, while this report will contain many new facts for geological readers, they must expect to find them stated in simple language. The limits imposed on the writer have permitted only a brief reference to each part of the subject, but it has been his endeavour to allot the space impartially to all. If, therefore, those who may be most interested in any one branch should find the description of it too short to satisfy them, they must consider the claims of all the others. It is hoped that, should the demand warrant it, a fuller report may be issued at a future time. **Technical terms.**

Owing to the uncertainty which has heretofore prevailed in reference to the northern boundary of the province, it will be necessary in attempting a geological description of Ontario to state at the outset how far we understand our territory to extend in that direction. For the purposes of description we will assume that the Albany river is the northern boundary all the way to the sea, and that a meridian line from James bay to the head of lake Temiscaming, and the Ottawa river thence to Point Fortune, constitute the eastern boundary. **Bounds of the territory.**

In order to facilitate our description and to prevent repetition we will here present a table, showing in their proper order all the divisions of the rocks of the province. [See next page.]

DIVISIONS OF THE ROCKS OF ONTARIO.

In descending order.

SYSTEM.	
	Recent.
Post Tertiary.	Soils, Peat, Shell-marl.
	Lacustrine and Fluviatile Clays, Sands, etc.
	Pleistocene.
	Saugeen Clay, Artemesia Gravel, Algoma Sand.
	Sand, Gravel and Shingle of the country north of the Great Lakes.
	Erie Clay, Calcareous and Non-calcareous Clays north of the Great Lakes.
	Boulder-clay, Drift or Till.
	Palæozoic.
Devonian	Chemung and Portage.
	Hamilton Formation.
	Corniferous Formation.
	Oriskany Formation.
Silurian	Lower Helderberg Formation.
	Onondaga Formation.
	Guelph Formation.
	Niagara Formation.
	Clinton Formation.
	Medina and Oneida Formation.
	Hudson River Formation.
	Utica Formation.
	Trenton Formation.
	Black River and Birds-eye Formation.
	Chazy Formation.
	Calciferous Formation.
Cambrian	Potsdam Formation.
	Nipigon Formation.
	Animikie Formation.
	Azoic or Archæan.
Huronian	Upper (?) Huronian Formation.
	Lower (?) Huronian Formation.
Laurentian....	Upper Laurentian Formation.
	Lower Laurentian Formation.

Igneous or eruptive rocks may be of any geological age, and those which occur in Ontario will be noticed in describing the systems or formations to which they are supposed to belong. In the list in the table the divisions of the rocks of Ontario are presented in their natural order of succession. It does not by any means represent the complete geological scale, comprising only the newest and some of the oldest systems. There is an enormous gap between the Post Tertiary and the Devonian, which in a complete section of the earth's crust would be filled up (in descending order) with the Tertiary, Cretaceous, Triassic, Permian and Carboniferous. The whole of the geological scale is not found in any one region of the surface of the earth, but the order of succession has been ascertained by tracing the connection of one with another, principally by the aid of the fossils or organic remains which they contain. Between the time of the disposition of the highest or newest of our Devonian rocks and the oldest of the Post Tertiary a vast interval elapsed, during which this part of the world may have been dry land and little or nothing may have been deposited upon it. But it is far more likely that rocks of some, at least, of the systems now wanting were laid down and have long since decayed and disappeared through the action of denuding agencies; while elsewhere the conditions have been more favorable for the preservation of some of them in one country and others in another. Gaps in the System.

In describing the rock-formations of Ontario we propose to begin at the bottom of the scale, or with the oldest, and proceed in the natural order or that of their age. First, however, a few words may be necessary in regard to the terms employed and the names of the divisions themselves. Geological divisions.

The term 'system' in geology is used to designate great series of strata characterised by such similarity that they may 'stand together,' as the word implies. In the Azoic or Archæan division the rocks themselves comprising a system have certain points of resemblance in common, while among the fossiliferous strata each system is recognised by the remains of some prevailing forms of animal or plant life. The systems are intermediate in comprehensiveness between the periods or ages and the formations, each system usually comprising several formations. System.

The 'formation' constitutes, as it were the unit, in the geological classification or grouping of the rocks. Among fossiliferous rocks each formation comprises strata which may be distinguished from all others by their organic forms, most of which are peculiar to such formation. Non-fossiliferous formations comprise rocks which have a recognised position in the scale, or which possess some strong points of resemblance sufficient to distinguish them; or they consist of rocks which have been formed under similar conditions and, as far as can be ascertained, at about the same time. Unfortunately the term formation has been employed by some geologists rather loosely, or without a uniform and definite signification, and of late years an attempt is being made to give it a more extended meaning, by which it would take the place of the well-established term 'system.' Formation.

The word 'group,' which is so often used in geological language, is another which does not yet enjoy a universally established meaning. Heretofore Canadian geologists have been accustomed to use it as intermediate Group.

in comprehensiveness between system and formation. Thus we spoke of the St John group, the Quebec group, the Trenton group, the Anticosti group, each embracing a number of formations. At the present time some European geologists are seeking to give the term a larger signification, equivalent to system, or even period.

But the word which has been used most loosely of all in geological language is 'series,' which is still made to do duty wherever there is any uncertainty as to the rank of any set of rocks.

Series.

In regard to the proper names for the various divisions of our rocks, the late Sir William Logan, when he undertook the geological survey of the province of Canada in 1842, wisely foresaw the advantage of adopting the names already in use in the state of New York, adjoining us. In this way there was no confusion, and everyone understood without further explanation the positions of our various formations as described by Logan under these names. Geological formations are distributed in the crust of the earth without reference to national boundaries, and true geologists are the most cosmopolitan of men, the whole earth being their field of research, as the very name 'geology' implies. The New York state and other American geologists had adopted the names for the Systems which had been given in England, such as Cambrian, Silurian, Devonian, Carboniferous, etc., but as the subdivisions of these or the formations in America could not be closely correlated with those of England, local names had to be adopted. Most of the formations of Upper and Lower Canada were found to be continuous with those of the adjoining states, so that the names for these were applicable on both sides of the international boundary line. In a few cases, such as that of the Hamilton formation, named after the village of Hamilton in Madison county, N. Y., some misconception has arisen from the supposition that the name is derived from our own city of Hamilton. Professor Chapman has proposed the alternative name Lambton formation, as it occurs chiefly in Lambton county in Ontario. One of our Ontario formations, the Guelph, is not represented in the state of New York, and the name which it bears was proposed for it in 1861 by the writer, after the city of Guelph, which is built upon it. The name Nipigon was also proposed by the writer for one of the formations of the lake Superior region, on account of its local importance and peculiarities, and because of a doubt as to its equivalency with any of the formations which had been already named.

Ontario names for Systems and Formations.

The Saugeen Clay, Artemesia Gravel, Algoma Sand and Erie Clay, the names of which were also proposed by the same geologist and adopted by Sir William Logan in the 'Geology of Canada,' constitute formations which are distinguished mainly by the characters of the deposits themselves, although organic remains have been found in some of them. The name Animikie, for an important formation on the north-west shore of lake Superior, was proposed by Dr. T. Sterry Hunt in 1871, two days before Dr. Bell had suggested Lower Nipigon for the same formation, and the former term has been retained. The terms Huronian and Laurentian were given by Logan and Hunt early in the history of the Geological Survey, and have been followed by geologists, not only for Canada, but in all quarters of the globe where rocks of corresponding

Systems exist. About the same time the name 'Lawrentian' was suggested by another geologist for the Post Tertiary clays and sands of Vermont and Lower Canada, but it was soon after dropped, these deposits becoming known as the Champlain clays and sands.

Other names for some of the systems and formations represented in Ontario have been more or less employed by geologists, and these will be mentioned in the more detailed descriptions to follow; but in order to preserve simplicity they have not been given in the table.

GEOGRAPHICAL DISTRIBUTION.

Geographical features of geological divisions.

Before proceeding with an account of the geological or lithological nature of each of our formations and of their economic minerals, we shall give a brief outline of the leading geographical features of the principal divisions. Along with the descriptions of the individual formations, the areas which they severally occupy will also be given in sufficient detail. The Recent and Pleistocene clays, sands, etc., are called superficial deposits, and the older and harder rocks below them in Ontario may be termed the fundamental rocks. The ordinary geological map of Canada represents the latter only, as if the superficial deposits did not exist. It would be difficult to show both at the same time, as these deposits are spread independently over all the older rocks alike. A separate map for the superficial deposits therefore became necessary, and such a map was prepared by the writer and published in the atlas which accompanied the 'Geology of Canada' in 1863.

Structure of the continent.

In order to form a clear idea of the general features of the geology of Ontario it will be desirable in our introductory remarks to go beyond the immediate borders of the province, and consider for a moment some points bearing on the structure of the continent.

The Hudson Bay slope.

The most northerly section of Ontario, or that bordering on the lower part of the Albany river and James bay, resembles the most southerly portion, or the peninsula between lake Huron and the lower lakes, in being underlaid by almost flat-lying Silurian and Devonian rocks, while the great intermediate tract is occupied by a part of the Azoic area which stretches to the Arctic regions. Most of this tract consists of Laurentian gneiss, but between lake Huron and James bay there is a very large district of Huronian rocks which are of much importance an account of the economic minerals they contain. The Palæozoic rocks coming within the province in the northern or James bay region occupy an area almost as great as those of the whole southern peninsula of the province, and as they extend beyond the Albany river their total area on the west side of James bay is much greater. In both regions they are quite undisturbed, except in a few local cases, and remain in the almost horizontal positions in which they were originally deposited at the bottom of the sea. This is owing to the fact that they have been protected from movement by the massive and unyielding Azoic rocks that form the foundations on which they lie.

Palæozoic strata of the east.

This protection has not been extended to the Palæozoic rocks of eastern Pennsylvania and the region lying east of a line running up the Hudson river, through lake Champlain, and thence to the city of Quebec and down the lower St. Lawerence. To the east of this line a mighty force, supposed to be due to the gradual shrinking of the earth, has acted for ages from a south-easterly direction, and has caused the great undulations in the strata that now form the Appalachians, the Green and White mountains and the Notre Dame range, extending into the Gaspé peninsula. It has also produced great dislocations or faults and overturnings of the strata. The Palæozoic rocks of both the northern and southern extremities of Ontario having been sheltered from this force, their structure and geographical distribution are simple and have been worked out with comparative ease.

Foldings of the Azoic rocks.

But the Azoic rocks are highly disturbed, and much more folded and contorted than the Palæozoic strata of the east. As a rule the foldings have been pressed together so completely that their anticlinals and synclinals have taken the form of sharp A's and V's, and the normal position of the stratification is usually more nearly vertical than horizontal. The lateral pressure which caused this has probably been due also to the shrinking of the whole globe. Besides folding the Azoic strata in the manner described, this pressure has developed in them a slaty cleavage, whenever their nature would permit of it, and has also aided in producing their prevailing crystalline texture. But all this took place before the formation of the Palæozoic strata, which rest almost horizontally upon the truncated edges of the crystalline rocks.

The Palæozoic region of southern Ontario.

The unaltered fossiliferous beds of southern Ontario form part of a great Palæozoic region that stretches over most of the northern states, while those of the northern extremity of the province appear, from their fossil evidence, to have been deposited in a part of the ancient sea which must have been separated from the main body, much as Hudson bay is now separated from the Atlantic ocean.

THE AZOIC PERIOD.

Azoic rocks of Ontario.

This great division is so called because, as yet, no trace of either animal or plant life has been found in it. It is also termed the Archæan period or age. In Ontario the rocks which belong to it may be grouped under the Laurentian and the Huronian systems, although other divisions have from time to time been proposed for some of them. These two divisions are considered sufficient by many geologists for the Azoic rocks of the whole world. Without taking local peculiarities into consideration, the primitive rocks of all countries may be classified under one or other of these great Systems, even if subordinate divisions should be found convenient in some localities. The characters and proportions of the different rocks which make up the Laurentian and Huronian are naturally found to vary much in different regions, although they are everywhere essentially the same Systems and retain the same relative positions, representing similar conditions in the geological history of the globe. They form the foundations of the crust of the earth as far as we can observe or penetrate it, and are easily separable from any rocks lying above them. Their crystalline characters and generally disturbed condition are their distinguishing features. At the same time it is true that, in some instances, newer rocks have been so altered locally or even over considerable tracts as to resemble the Azoic, but we generally find some means of distinguishing between them. In Canada and the United States the Laurentian and Huronian are usually intimately associated, but their lithological features, or the internal characters which distinguish rocks from one another, are sufficiently distinct to separate them. As they are for the most part included in one great area they must be to some extent described together.

The Azoic rocks of Canada have been represented as extending from the region of the great lakes in the form of two arms, one stretching north-eastward to the Atlantic coast of the Labrador peninsula and the other north-westward to the Arctic sea, east of the month of the Mackenzie river, the intervening space being filled up with Palæozoic rocks. Further light on the subject has, however, shown that the geographical outline of these rocks takes the form, approximately, of an immense ellipse which includes the north-eastern part of the continent, Baffinland, Greenland and many of the islands of the frozen sea. It comprises the whole of the Labrador peninsula, measuring a thousand miles each way. On the other side its boundary runs, with a westward curve, from lake Winnipeg to Coronation gulf, another thousand miles, with a spur towards the mouth of Mackenzie river. The Palæozoic rocks of Hudson bay form a sort of broken fringe around that inland sea, and a belt of them extends thence northward across some of the islands to the Arctic ocean. The geographical depression of Hudson bay, to which the rivers flow from all sides, forms the central drainage basin of this Azoic area of North America, and its origin is of very ancient geological date. At various periods of the earth's history it was probably covered by waters more or less separated from the outer ocean, and the newer rocks in its centre were deposited from these in the same way that deposits are forming in the bottom of the bay at the present time.

Geographical distribution of Azoic rocks.

General outline.

Although the superficial continuity of the Azoic region just described is broken in many places by channels of the sea, and by outlying patches of Palæozoic rocks, it may be regarded as practically one area of compact outline, and it forms the nucleus upon which the rest of the continent has been built. On the east it falls abruptly into the deep ocean, but on its landward sides it is flanked by the formations which have been successively deposited around it. The further we recede from it the newer the rocks become, till in one direction we reach the Rocky mountains, which have broken up through a vast thickness of these succeeding strata.

Nucleus of the continent.

As a rule the Huronian rocks are less contorted or corrugated on the small scale than the Laurentian, but on the large scale they partake of the same foldings which have affected the latter. At one time they were supposed to be less abruptly bent into anticlinal and synclinal forms, but this appears to have been a misconception, due to the fact that some of the highest beds happened to have been first studied in a district that is less disturbed than the average. In other localities some of the Laurentian rocks are quite as little disturbed.

The greater part of the mixed Laurentian and Huronian region belongs to the former, and of it, the Lower Laurentian is the prevailing type. As represented on a map, the Huronian occurs in the midst of the Laurentian in the form of more or less completely separated areas, or with straggling connections between them. They seem to be in a manner interwoven with the Laurentian as basins or troughs more or less elongated, and as tracts of angular and other forms filling spaces between great nuclei or rounded areas of Laurentian rocks. Patches of Huronian strata of com-

A region of mixed Laurentian and Huronian rocks.

paratively small size are numerous throughout this vast Azoic region of the north-eastern part of the continent, and in addition to these there are a few of great extent. One of them is on the north-west side of Hudson bay, and appears to stretch far inland. Another lies to the north and north-east of lake Huron, reaching from the east end of lake Superior almost to lake Mistassini, a distance of 600 miles. In Wisconsin and Michigan also considerable areas exist, and in the country between lake Superior and lake Winnipeg, Huronian rocks of many different basins are largely mixed with the Laurentian, constituting perhaps one-third of the whole area. In the country between the northern extremity of lake Winnipeg and Hudson bay the writer has described a Huronian trough 180 miles in length, and Mr. A. S. Cochrane found these rocks between the Saskatchewan and Churchill rivers and largely developed on the north side of lake Athabasca.

THE LAURENTIAN SYSTEM.

We have given the above brief account of the relations of the Laurentian and Huronian systems to each other, and of the distribution of the two in north-eastern America, in order that the reader may be better understand what is to follow in regard to the rocks that occupy the greater part of Ontario as now extended. The country formed by these two systems is sometimes referred to as the Laurentian region, but it is more correctly called the Azoic or Archæan when areas of both classes of rock are included. We shall now proceed with a short description of the Laurentian alone.

Lower Laurentian or Primitive Gneiss series.

As indicated in the table already given, the Laurentian system has been divided into two formations, the lower of which is sometimes also called the Primitive Gneiss series. The differences between them can be best pointed out after having described the Lower Laurentian. Both formations give rise to the same kind of country which is so familiar to all Canadians. As a rule it is hilly, but not greatly elevated above the sea, and full of lakes. Within the regions which have been sufficiently explored to speak of with some degree of certainty these amount literally to tens of thousands, and occupy a very considerable proportion of the whole surface, estimated in some sections at one-third and even one-half of the whole area. The cause of the existence of these lakes will be explained further on. The high northern part of the coast range of eastern Labrador has not been glaciated, but almost everywhere else there are unmistakable signs of this phenomenon. This has given rise to the peculiarity of the Laurentian country which Sir William Logan has so graphically described as mammillated. This vast hilly country, however, cannot properly be called "the Laurentian range."

LOWER LAURENTIAN FORMATION.

Character of the Lower Laurentian.

The Lower Laurentian consists essentially of gneiss. In some localities its foliated or stratiform character is obscure, and it may be called granitic or syenitic. The distinctly banded varieties differ from one another considerably in the proportions of their constituents. Typical gneiss is defined by lithologists to consist of quartz, felspar (orthoclase) and mica, but most of the gneisses of both the Lower and Upper Laurentian contain hornblende, often in large proportion. These would be called hornblendic or syenitic gneisses. The proportions of these minerals vary constantly, and it is seldom that there is any great thickness having the same composition. One layer may consist chiefly of felspar and quartz, the next may contain much hornblende or mica in addition, while a third may consist largely, of any one of these alone. These minerals, in fact, enter into the composition of all the gneissoid rocks in every conceivable proportion. It is easy for the mere lithologist to select typical varieties of rocks in a good cabinet collection, but in the case of the gneissoid rocks it is impossible for the field geologist to recognise these distinctions on a large scale. In the Lower Laurentian, hornblende is almost as generally diffused as the felspar, quartz and mica. It sometimes occurs as bands consisting principally of this mineral in both the lower and upper divisions. In the latter it has been noticed particularly in proximity to the limestone bands and the iron ore deposits. The Laurentian hornblende rocks are usually blacker and more coarsely crystalline than those of the Huronian system.

Varieties of gneiss.

The prevailing colors of the Lower Laurentian gneisses are greyish and reddish, from very light to very dark shades, depending partly on the colors and partly on the proportions of the different constituents. The felspar (orthoclase) is white, grey and red, or sometimes yellowish or greenish; the quartz is white to grey, and the mica and hornblende black, or very dark green or brown. These rocks are generally distinctly foliated, or show a lamination or parallelism in the arrangement of their constituent minerals easily traceable by their colors. Where the latter are very distinct and the layers continuous and close together, the rock in cross-section is described as ribboned; where the layers are further apart it is called banded. But the bars are often broken into a series of tapering dashes which pass below or above each other, or with an interlocking or "dovetail" arrangement, or the bars may be connected by thin streaks or rows of dots. Even where the tendency to parallelism in the texture of gneiss is not conspicuous, from the want of contrast in colors, it can always be seen on close inspection, and this kind of structure or "grain," which may be compared to that of wood, is what distinguishes gneiss from granite, the latter having no such parallelism in the arrangement of its constituent minerals. On the supposition that this structure of gneiss, even when the parallel bands of different kinds are quite thick, may be accounted for in other ways than by stratification originally due to the action of water, some geologists hesitate to speak of it as stratification or bedding, notwithstanding its apparent identity with it.

Color and form of gneiss rocks.

As a rule, in Canada the exposed surfaces of the gneiss rocks show little sign of decay, on account of their having been worn down by glaciers in comparatively recent geological times, and they have an extremely massive appearance. When broken up, as by blasting, they fracture almost impartially in all directions, or show only a slight tendency to cleavage along the planes of their foliation. This foliation in the gneisses of the Lower Laurentian is usually contorted or bent in various directions on the small scale, and any differences in their composition or color do not appear to be sufficiently persistent to trace them far in any direction on the ground; in other words, they are not sufficiently differentiated into great bands of distinct kinds as to enable them to be shown on a map of moderate scale, as is often the case with the gneisses and other rocks of the Upper Laurentian. Still, in those areas which

Foliation and strike.

have been most examined, a general tendency has been observed to strike more nearly in a north-easterly and south-westerly direction than in any other. In eastern Labrador, and also in Baffinland, the larger mountain ridges run north-westward, but it has not been ascertained that this is the direction of the strike of the gneiss in those regions. The monotonous grey and red massive and contorted gneiss above described prevails throughout the vast Lower Laurentian region, stretching from the great lakes of the St. Lawrence to Hudson bay and thence to lake Winnipeg, as well as in the western and most of the southern parts of the Labrador peninsula.

Dykes and veins in the Laurentian system.

In some districts the Laurentian rocks are cut by dykes of greenstone or trap, many of them very large and affecting the geographical features. Rivers or long narrow lakes sometimes lie upon the courses of dykes which had become decomposed and yielded to glacial action, while falls and rapids occur where hard dykes cross the courses of streams. Both the Lower and Upper Laurentian formations are cut by veins of two classes, the first being much more ancient than the second. The former, which are numerous, are, as it were, fused into or amalgamated with the country rock and are composed of the same minerals. In some cases the gangue is almost entirely felspar, in others quartz, but oftener the two minerals are mixed together and a little mica or hornblende is added. The larger veins of this class are very coarsely crystalline; the smaller ones have a tendency to branch off or become reticulated. Although the division between them and the wall-rock is distinctly defined by the contrast of colors, there is no actual separation between them, the two breaking like one rock. Metallic ores have not been found in these veins in economic quantities. Veins of this class may be seen in almost any locality where the gneisses are exposed. The veins of the second class are not so common, and have been formed long subsequent to those of the first class. Their gangue, which is frequently calcspar, separates easily from the wall rock, and is apt to contain galena, copper and iron pyrites and zinc blende; but these minerals, like the veinstones themselves, have perhaps been derived from rocks resting on the gneiss, or which rested upon it at some former period when these veins were formed, but which have since been removed by denudation. The lead-bearing veins of the counties of Frontenac and Leeds, and those north of the Canadian Pacific railway opposite the head of Black bay, lake Superior, are examples of the second class. With the exception of the contents of veins of this class and the coarsely crystalline felspar and quartz of those of the first class, no minerals of economic value are known to occur in Canada in the Lower Laurentian formation or primitive gneiss series above described.

UPPER LAURENTIAN FORMATION.

Question of conformity.

Under this name Sir William Logan described a series of massive labradorite and anorthosite rocks, such as those north of Montreal, in the region of the upper Saugenay, on the north side of the St. Lawrence just below Quebec, and on the Moisie river; and similar rocks are found on the west side of lake Champlain. He thought that they might be above and unconformable to the gneisses, or the interstratified gneisses and limestone nearest to them. Professor James Hall agrees with Logan's view. Dr. Selwyn thinks they may be interstratified with the gneisses and limestones. In Parry Sound district, where the writer found anorthosite rocks, they are interstratified with gneisses, etc., with which limestones, similar to those of the county of Argenteuil, in Quebec, are also associated. It would appear from the writer's observations over the vast Laurentian regions of Canada that for the present, at least, it will be convenient to designate as Upper Laurentian both the anorthosite rocks such as those above referred to, and the limestone-bearing series such as that which was so carefully worked out by Sir William Logan in the county of Argenteuil and sometimes called the Grenville series, as there are good reasons for this classification, and it is the most convenient one in the present state of our knowledge. In the counties of Terrebonne, Montcalm and Joliette, in Quebec, rocks similar to the Grenville series have been found since Logan's time to be interstratified with anorthosite gneisses. Dr. Selwyn regards the more massive anorthosites or labradorite rocks of Argenteuil, Terrebonne, etc., as probably of igneous origin, and as in some way incorporated with the adjacent limestone-bearing series as already stated. Professor Hall, the state geologist of New York, considers the similar rocks, which are largely developed on the west side of lake Champlain, to overlie the adjoining gneisses unconformably. Both views may be correct. It is highly probable that volcanic activity went on more energetically and on a grander scale in these early days of the earth's history than now, and great outbursts of basic matter, such as these anorthosites, were of frequent occurrence in Laurentian times. After spreading out upon the surface of the earth or on the bottom of the sea, some of them became incorporated in a conformable manner with the contemporaneous deposits, while others may have flowed over pre-existing rocks which were even then consolidated and disturbed. The latter would form the unconformable masses of Logan and Hall.

Mineral-bearing character of the upper series.

There may be a general want of conformity between the primitive gneisses of the Lower Laurentian and all the rocks of the upper series which succeed them, including both the massive anorthosites and the limestones with their accompanying gneisses. There is a considerable change of character in passing from one to the other, and this is important from an economic point of view. While the lower Laurentian is apparently barren of metallic ores, the upper series, as above defined, contains a considerable variety of them. In addition to the presence of the limestone and dolomite bands and the anorthosite rocks which constitute their leading distinguishing features, the Upper Laurentian is characterised by the occurrence of iron ores, graphite, apatite, pyroxene and hypersthene rocks, quartzite and argillite bands, granite, syenite and porphyry, and perhaps conglomerates. Besides the above differences, Mr. W. C. Willimott enumerates the following sixty-one species of minerals as found more or less commonly distributed in these rocks, in Canada, few or none of which have yet been detected in the Lower Laurentian:

Upper Laurentian minerals.

Achroite.
Actinolite.
Agate.
Allanite.
Amazon-stone.
Anorthite.
Aphrodite.
Aventurine felspar.
Axinite.
Barite.
Beryl.
Bismuthinite.
Bismuth carbonate.
Blende.
Bornite.
Celestite.

Chabazite.	Oligoclase.
Chrondodite.	Peristerite.
Chrome garnet.	Perthite.
Chromite.	Picrolite.
Chrysotile.	Pyrallolite.
Corundum (Hunt).	Pyroxene (and its varieties, sahlite, diopside and cocolite).
Essonite.	
Fluorite.	Pyrrhotite.
Gold.	Rutile.
Graphite.	Samarskite.
Heulandite.	Scapolite.
Idocrase.	Serpentine.
Jasper.	Steatite.
Labradorite.	Sphene.
Limonite.	Spinel.
Malachite.	Stilbite.
Microcline.	Talc.
Mispickel.	Meneghinite.
Molybdenite.	Tourmaline.
Molybdite.	Tremolite.
Monazite.	Uranite.
Mountain cork.	Zircon.
Nickeliferous pyrites.	Zoisite.

Boundry of the Upper Laurentian series.

No attempt has yet been made to mark the geographical division between the Lower and Upper Laurentian as above defined. But if we draw a line from near the north shore of lake Nipissing north-eastward, nearly parallel to the St. Lawrence, and at an average distance of about 150 miles from it, we may not be far from its position, for the region between lake Huron and the Gulf of St. Lawrence.

Supposed aqueous origin of gneiss in the Upper Laurentian.

Notwithstanding the various differences which distinguish the Upper from the Lower Laurentian, there is often a close resemblence in the gneisses of the one to those of the other. In fact it would often be difficult to distinguish between hand specimens taken from the two series. This fact should be borne in mind in considering the origin of gneisses in general. As the balance of evidence is strongly in favor of the aqueous theory of the origin of at least part of the Upper Laurentian, this lends support to the view that even the primitive gneisses may have been formed by the action of water during some early condition of the earth, of which we can form but little conception, judging by the later stages of its history. Minute globules of water have been found by microscopic examination in the centres of crystalline grains forming gneiss, and more or less water may be driven out of these rocks by means of heat.

The Eozoon Canadense a scientific myth.

At one time, some geologists alleged that they had detected an organic form, to which they gave the name of "Eozoon," in the Upper Laurentian rocks; but on investigation by others the hypothetical discovery was not endorsed, and the organic nature of the supposed fossil has been repudiated by nearly all scientists. It is believed that organic life not only did not begin on our planet in Laurentian times, but that for ages afterwards the earth was not fitted for its reception. Forms like the branching structures which are the portions of the so-called eozoon, said to have a resemblence to certain organisms, are assumed by a great variety of minerals, but in the case of eozoon these forms are unlike any organic structure in the fact that they are al different one from another.

It has been asserted that the limestones, iron ores, graphite and apatite are also evidence of the existence of animals or plants in Laurentian times. Such an argument, however, appears to have no good foundation. The limestones have been carefully examined by numerous geologists over immense regions during the last forty or fifty years and have yielded no evidence to support this view, but rather the opposite, namely, that they are of chemical origin. The iron ores occur in greater masses than any of those deposits which appear to have been aided by organisms in their formation, and, besides, their modes of occurrence are opposed to any theory of this kind. The graphite and apatite occur principally as vein matter. The largest deposits of the latter have been derived from the pyroxene rocks, which are evidently of igneous origin. Apatite is a common constitutent of traps and granites of all kinds, and is widely diffused in small grains even in gneiss. The argument as to its organic origin is based on the fact that phosphate of lime is the principal constituent of the bones of vertebrate animals; but, in the natural order of things, the phosphate must have existed first and the vertebrate animals later on. The converse of this is absurd. There is no evidence to show that phosphorus, carbon, iron and calcium did not enter into the original constitution of the earth as well as the other elements.

Origin of Laurentian limestones and iron ores.

ORIGIN OF LAURENTIAN ROCKS.

It is rather singular that the numerous minerals above referred to should appear for the first time in the upper Laurentian rocks, and this fact also suggests many questions as to their origin and mode of formation. One of the most remarkable features of the Lower Laurentian is the general uniformity in composition and character of the gneiss over so vast an area, almost semi-continental in extent. The comparatively fine-grained and even texture which prevails everywhere might not have been looked for in these ancient rocks, which have had such ample time to become coarsely crystalline or to exhibit local varieties. These circumstances seem to prove a uniformity of origin as well as of the condition of the surface of the earth at the time they were formed, as similar rocks also crop up from under all others in various parts of the world.

Origin and mode of formation of the Laurentian rocks.

Without stopping to consider the great differences in the various classes of rocks belonging to the Upper Laurentian, some geologists have suggested a general igneous origin for the whole of them. On this hypothesis it is supposed that these rocks may be compared to slags which have formed on the surface of a molten mass, and that instead of cooling quietly and homogeneously some force has acted on them, giving a sort of flow-structure such as may be seen in slags which have run from a blast furnace. It is difficult, however, to conceive how such a minute and even structure could have extended through so great a thickness, the depth of which is unknown, but which must amount to many miles in the Lower Laurentian alone. Others attempt to account for the stratiform condition of the Laurentian rocks by the agency of pressure, but it is impossible to imagine that this force could separate the rocks into immense bands of different kinds such as those of the Upper Laurentian, which differ from each other as much and on as large a scale as do the rocks of any of the later Systems. When we consider the

Their supposed igneous origin.

great variety and thickness of the Upper Laurentian, amounting to 50,000 feet or more, and including great bands of different sorts of limestones, dolomites and gneisses, and smaller ones of schists or quartzites, argillites, bedded iron ores, etc., the conclusion appears irresistible that they are, principally at least, altered sedimentary strata, as Logan so emphatically stated after having studied them for years in the field. The circumstance that many varieties of the Upper Laurentian gneisses are undistinguishable from those of the lower series would indicate that, like the latter, they have been formed during a primitive and probably heated condition of the earth. Sir William Logan's painstaking investigations in the county of Argenteuil and other localities in the province of Quebec show the various thick belts, which he traced out in all their sinuosities, to conform in their geographical distribution to the structural laws of ordinary stratified rocks, where these have been thrown into undulations, or anticlinals and synclinals, and afterwards denuded.

Altered sedimentary strata.

The Upper Laurentian rocks seem to be much more limited in geographical extent than the lower. Including both the anorthosite and the limestone-bearing portions under this designation as above described, the series, as already stated, may be said to extend from the lower St. Lawrence westward to lake Huron and northward in some places for about 150 miles, but beyond that distance it has not been recognised. It is largely developed between the Ottawa river and the Palæozoic region north of lake Ontario, and again between Georgian bay and lake Nipissing. Some of the rocks of the Hastings and Lanark region which were formerly included in this series are now believed by certain geologists to belong rather to the Huronian

Extent of the Upper Laurentian series.

Anorthosite or labradorite rocks are extensively developed in eastern Labrador, and there are indications in that region of iron and other ores as well as of non-metallic minerals indicative of the upper series. On the shores of some parts of Hudson straits the gneisses are divided into great bands with distinctive characters, and here iron ore, iron pyrites, graphite, sphene, sheet mica and some of the other minerals of this series are found, and in one place the writer noticed a loose piece of crystalline limestone of one of the varieties peculiar to the Upper Laurentian.

Rocks of Labrador.

The Laurentian and to an almost equal extent the Huronian districts of Canada are characterised by great numbers of lakes of all sizes, and mostly of very irregular forms. Throughout these vast but little explored regions they exist in tens of thousands It is estimated that, in some sections of this land of lakes, from one-third to one-half of the entire area is water. Some of them are one hundred miles and upwards in length, and many measure from twenty to fifty miles. They generally show a tendency to run in chains or groups in different courses, and thus they afford a means of travelling by canoe in almost any direction. They are nearly all rock-basins, and on the water-sheds they not infrequently have outlets in opposite directions, sending the water down either slope. One of the most remarkable examples of this phenomenon in Ontario occurs on the divide to the north-east of lake Nipigon, where rivers of equal size flow from each end of Summit lake, one into Hudson bay and the other into the St. Lawrence. Both

A land of lakes.

Lakes with double outlets.

streams are uninterruptedly navigable by canoes for some distance after leaving the lake, so that travellers may here cross the water-shed by continuous navigation.* Temagami lake, north of lake Nipissing, one of the most picturesque lakes in America and measuring thirty miles each way, is another striking example of this kind. Its northern outlet flows into the Ottawa by way of the Montreal river and its southern into the St. Lawrence by Sturgeon and French rivers. What is still more remarkable—this lake, as shown by the writer's explorations, had in former times an eastern outlet to the Ottawa and a western to the St. Lawrence, and if its level were now raised only a few feet these channels would again help to drain it, so that it might have four outlets flowing towards all the cardinal points of the compass, the two pairs of opposite discharges being each about thirty miles apart. No fewer than eight examples of lakes with double outlets are known among the upper branches of the Ottawa.

Ontario a group of islands.

The question naturally arises—What was the probable origin of these innumerable lakes? Their formation and that of the boulder clay, and the existence of the surface boulders consisting of primitive rocks that are so abundant over the Azoic regions and often beyond them, as well as the rounding, fluting and grooving of the rocks, which may be seen almost anywhere on the removal of their superficial coverings, are all mutually related phenomena. Without going into details, it may be stated that before the beginning of the glacial epoch the surface had been subjected to a long period of decay. Under the influence of water and other decomposing agencies the crystalline rocks became softened to a great depth, as may now be seen in similar rocks in South America and elsewhere. One of the best examples of this in Canada is to be found in the north-eastern part of the Labrador peninsula, where the same gneisses, etc., which in Ontario we are accustomed to see as hard rounded hills and knobs, form pointed mountains and ridges in a soft and decayed condition so unlike the former as to be quite unrecognisable at first sight. When the cold of the glacial epoch set in, ice formed to a considerable thickness over all the elevated parts of the land and descended in wide sheets in various directions upon the lower levels, carrying with it the softened or decomposed rock, which by the weight and motion of the ice became ground up to form the finer parts of the boulder clay or till. From the subsequent washing away of portions of this by the action of the sea and glacial lakes the Pleistocene clays and sands have been deposited. The glacial action continued till all the decayed crust had been carried away, and the bottoms of the glaciers were grinding on the sound rock beneath. What would we expect as the result of these processes? Naturally when the glacial ice disappeared, through a change of climate, there would be a hard, rounded but uneven surface, as the rotting of these crystalline rocks had taken place to unvarying depths according to the previous contour of the land, or to the more or less decomposable nature of the different bands and in proportion to the unequal hardness of the undecomposed parts after they had been reached by the powerfully eroding glaciers. The angles of the dips, the curves and twists in the strike, the joints, and par-

Glacial origin of the lakes.

* See Geological Survey Report by Dr. Bell for 1871.

ticularly the dykes cutting these rocks, would all have their effect in modifying the form of the surface left at the close of the glacial epoch. The relative levels of the land, except on a continental scale, and its general contours, including the principal river valleys, probably existed before this epoch, and the latter guided the courses of the diminishing glaciers towards the close of the period, as may be seen by the direction of the striæ on the rocks along their bottoms. On the disappearance of the ice, the innumerable rock-basins which had been formed by the above process would be already full of water, and the rivers running in their present channels.

Origin of boulders.

It has been observed that in the decaying of crystalline rocks, as above described, small portions here and there remain unaffected, forming as it were kernels in the mass. These, with a little rounding by the action of the moving glaciers, would become the boulders or "hard-heads," so common in most parts of the country.

Decay of rocks still going on.

The high inclinations and almost vertical attitudes of the bedding or cleavage in most of the crystalline rocks has greatly favored the penetration of water and air, and their consequent destruction. In excavating the comparatively sound rock left under the glaciated surface at the present day, as in quarries and railway cuttings, we see how deeply the effects of these agencies are traceable, and they are even now laying the foundations of further decay.

The above short sketch of some of the most important points bearing on the superficial geology of Ontario is almost inseparably connected with the description of the geographical distribution and peculiarities of our Laurentian and Huronian country, and it will serve for reference in noticing the Pleistocene deposits in a subsequent part of this report.

Economic minerals.

The economic minerals of the Laurentian system and their modes of occurrence will be alluded to in the general description of the various kinds, under their respective headings further on.

THE HURONIAN SYSTEM.

Characteristics of the Huronian rocks.

This is the second principal division of the rocks of the earth in ascending order. In Canada it consists of a great thickness and variety of strata, for the most part crystalline, but in a less degree than the Laurentian, together with many unstratified igneous masses. Like the Laurentian it is azoic, or devoid of any trace of organic life, so that the distinction between the two systems is based entirely on lithological grounds. The difference in this respect is great, and is easily recognised by those who have paid any attention to geology. The prevailing dark green and grey colors of the Huronian offer a marked contrast to the lighter greys and reddish greys of the Laurentian. The latter are massive and coarsely crystalline, while the former are usually fine-grained and schistose or fissile, this cleavage structure constituting a striking difference from the solid Laurentian. There are some exceptions to this rule, such as the light colored quartzites and the granites and syenites of the Huronian to be noticed further on. The change in passing from one to the other is often sudden and complete, but sometimes beds of passage are met with.

Huronian areas

As already stated, most of the Huronian areas occur within the general boundaries of the Azoic or Archæan rocks, and their relations to the Laurentian have been pointed out. In addition to the numerous Huronian areas north of the great lakes and between Hudson bay and the North-west territories, there are certain rocks in the district stretching between the counties of Hastings and Lanark which may be regarded as belonging to this system. Some of the crystalline rocks of the Eastern Townships and of the provinces of New Brunswick, Nova Scotia, especially on the south-east side of Cape Breton, and Newfoundland also appear to belong to this division. The writer has found small Huronian areas on the eastern shores of James and Hudson bay's, and in the eastern part of Labrador.

Disturbances.

In the midst of the larger Huronian areas limited portions are found where the strata are less disturbed than usual, or dip at moderate angles, just as similar conditions are occasionally found among the Laurentian rocks; but a local circumstance of this kind does not affect the general character, nor afford a reason for separating these portions of either system from the rest. Examples of little disturbed sections of these rocks are met with near the Bruce Mines, lake Huron, on Montreal river and lake Temiscaming. Although the Huronian strata have generally been thrown into sharp folds, or stand at high angles, they are as a rule less bent about or contorted than the Laurentian.

Volume of the system.

In order to ascertain more correctly the nature of the Huronian deposits, the writer, in connection with the Geological Survey, commenced a detailed investigation of the Lake-of-the-Woods area in 1881 and 1883. These researches have since been continued by Dr. A. C. Lawson, and have shown that the different bands are not often persistent for any great distance, and that they vary much in thickness and change in lithological character when followed out on their courses. It therefore becomes difficult to estimate their thickness, even in a given area. From the sections measured by the late Mr. Alexander Murray, principally among the higher members of the series behind the Bruce Mines, that gentleman calculated that they have there a volume of about 18,000 feet. The total volume of the system must be very great—probably not far from 40,000 or 50,000 feet, or perhaps even more.

Conformity of the Laurentian and Huronian rocks.

In Canada, as far as our investigations have gone, the two systems appear to be everywhere conformable to one another; but in rocks of such ancient date and which have undergone such profound structural changes, owing to pressure, etc., affecting alike the stratified and unstratified portions, this appearance may not everywhere indicate a truly conformable sequence. The manner in which the Huronian rocks occupy spaces with elongated or even angular outlines in the midst of the Laurentian areas has been already referred to. Both sets of rocks having been thrown by lateral pressure into sharp folds, standing at high angles to the horizon, the Huronian often appear to dip under the older Laurentian, but this is merely the effect of overturning, and does not show that a part of the Laurentian is newer than the locally underlying Huronian. Notwithstanding the geographical relations of the two sets of rocks, their general difference in character and composition would indicate that some great change in terrestrial conditions had occurred when the formation of the one system ended and that of the other began. In the Laurentian an "acid" or silicious composition prevails, whereas the Huronian rocks as a whole are more basic, chemically speaking. The latter can be shown to be very largely of volcanic origin, although this may not always be obvious at first sight.

Origin of the name.

The name Huronian (derived from lake Huron) was first given by the officers of the Geological Survey of Canada more than forty years ago, and it has been adopted by geologists in other countries as universally as the term Laurentian, and is made to include all the rocks lying between the Laurentian below it and the Cambrian or earliest fossiliferous rocks above. It thus forms an important and a convenient series, and its position in the geological scale is easy to recognise.

HURONIAN AREAS IN ONTARIO.

Huronian districts defined.

The greatest of all our Huronian areas forms a wide belt extending from the south-eastern extremity of lake Superior eastward along the north shore of lake Huron, from which it runs north-eastward, widening out till it occupies the whole country between lake Temiscaming and the head waters of the Montreal river, a breadth of one hundred miles. Beyond this it stretches north-westward across all the branches of the Moose river, northward beyond lake Abittibi, and north-eastward almost to the southern extremity of lake Mistassini, a distance of over 600 miles from the outlet of lake Superior. The Huronian area along the Ground-hog river, and Mattagami lake on its course, appears to be more or less completely separated from the great area above described. The next important Huronian district lies around Michipicoten at the north-east angle of lake Superior, running for sixty miles west and twenty miles south of that point, and extending inland to Dog lake, a distance of forty-five miles. Another large area stretches from the Pic river eastward or inland to Nottamasagami lake, and westward mingled with granites and green-stones, to Nipigon bay. Two extensive belts run eastward from lake Nipigon, one of which crosses Long lake. West of Thunder bay, and stretching to the international boundary line, there is a large area which gives off arms to the north-east and south-west; and several belts and compact and straggling areas occur between this and the Lake-of-the-Woods basin, one of which follows the course of the Seine river. The Lake-of-the-Woods area, which has been already alluded to, occupies the whole breadth of the northern division of that lake. An important belt starts between Rainy lake and Lake-of-the-Woods, and running north-eastward has a breadth of forty-five miles where it crosses the line of the Canadian Pacific railway. Minnietakie and Sturgeon lakes lie within this belt. Huronian rocks occur at both ends of lake St. Joseph and along three sections of the Albany river, between it and the commencement of the Palæozoic basin of James bay. The probability that some of the rocks of the Hastings and Lanark region may be classed with the Huronian has been already mentioned. The rocks of these various areas, and of others beyond the limits of Ontario, in some cases show considerable variations in the proportions of the different kinds of which they are made up. This is only what we would naturally expect where their origin has been local, as shown by the rapidly changing volumes of their different components, although they may have been nearly or quite contemporaneous. In one area steatites, serpentines and dolomites are abundant; in another, conglomerates, breccias

Local variations

and some amygdaloids; while in others we may find various crystalline schists, cherty and argillaceous slates, or it may be greywackés, quartz-diorites and quartzites, with slates and conglomerates.

LOWER AND UPPER DIVISIONS.

Although it will be difficult or impossible to draw any precise line of division applicable in all cases between the lower and upper parts of the Huronian rocks in Canada, yet for the sake of convenience they may be found separable, in a general way, into a lower and an upper series or formation. No horizon has yet been agreed upon at which to draw the line even locally, and the difficulty of an exact definition of a line is increased by the rapid changes of character in the lateral extension of any portion of the series. One of the United States geologists imagined that there was a general want of conformity between the lower and upper parts of what we had always called the Huronian rocks, and that a "basal conglomerate" was to be found at the contact of the two divisions; but there is no evidence whatever to bear out this supposition. Conglomerates are found indifferently throughout both the lower and upper portions. It may be a long time before we shall have worked out these rocks sufficiently to enable us to represent them separately on the map, but in the meantime it may perhaps be found convenient to speak of one set in a general way as distinguished from the other. Dr. Lawson proposed to call the Huronian rocks of the Lake-of-the-Woods region the "Keewatin" (more correctly spelled Kewaitin) series. This has the advantage of being a shorter name than the Huronian series of the Lake-of-the-Woods, but there are scores of other Huronian areas within the Dominion which are equally deserving of local names. Looking at the general geological map of the Dominion and the northern states, the Huronian as a whole is seen to occupy always the same place relatively to the rocks below and above it, and the general equivalency of all these areas cannot be disputed.

Undefined lines of difference.

The lower division consists largely of a variety of crystalline schists, in which the prevailing color is dark green or grey. Among these may be enumerated micaceous, dioritic, chloritic, argillaceous, hornblendic, talcoid, felsitic, epidotic, siliceous, dolomitic and plumbagenous. There are also crystalline diorites or diabases of various shades of grey and greenish grey (mostly dark), argillaceous and dioritic slate-conglomerates, granites and syenites, impure, banded and schistose iron ores, dolomites and imperfect gneisses. Among the commoner of the rocks of this division are fine-grained mica-schists, and dark-green dioritic or hornblendic schists. Two kinds of conglomerates are also abundant, one having an argillaceous matrix, with rounded pebbles of syenite and granite of various kinds and of some of the other Huronian rocks, but very seldom of gneiss; the other with a dioritic matrix, and often with rounded pebbles also. But in perhaps the majority of cases what were formerly considered as pebbles are really concretions of a lenticular form, and differing but slightly from the matrix in color and composition. They are best seen on wetted surfaces of cross sections of the rock, where they appear as parallel elongated patches tapering to a point at each end. Both hematite and magnetic iron ores are common in these rocks, and they

Peculiarities of the lower division.

have been largely worked in the Marquette and Republic districts in Michigan, and at Tower in northern Minnesota; but it is only lately that rich and workable deposits have been found among them in Ontario, although poorer ones have long been known in several localities. On the Antler river, about 100 miles west of Thunder bay, there is a very heavy deposit of rich and pure magnetite. Another is reported near the mouth of the Seine, and an extensive deposit of leaner ore to the east of Wabigoon lake. The late Professor R. D. Irving considered this feature of the Huronian so important that, for a short definition, he called it "a detrital iron-bearing series." But while the iron ores belong to the lower division, he attempted to restrict the name Huronian to some of the upper portions, which are not notably iron-bearing. The iron ores, whether workable or not, are generally accompanied by much red and dark jasper in thin layers. Gneiss is not common in the Huronian, and it differs from ordinary Laurentian gneiss in being imperfect, and also in being invariably slightly calcareous in all the numerous cases which have been tried by the writer. In some instances the felspar in it has been noticed to be triclinic, like those of the Upper Laurentian. Although rocks such as have been described as belonging to the lower division are largely developed in the Huronian areas to the west and north of lake Superior, they are by no means confined to these areas, but are met with in abundance in many parts of the great Huronian area north of lake Huron and elsewhere.

An iron-bearing series.

In the upper division probably the most abundant rock in Ontario is what may be called a graywacké, but which in the older reports was often styled a "slate-conglomerate;" but it also includes clay-slates, argillites, felsites, quartzites, ordinary conglomerates, jasper conglomerates, breccias, dolomites, serpentine, etc. In some localities the nearly vertical bands of quartzite, having withstood denudation better than the other rocks, remain as conspicuous hills or ridges, and this circumstance has caused their relative volume in the series to be over-rated by superficial observers. Within the province of Ontario these quartzites are most strongly developed near the height-of-land between lakes Abittibi and Temiscaming, and from the latter lake westward to the headwaters of the Montreal river. They are also common in the belt along the north shore of lake Huron, especially in its eastern part. The greywacké so abundant in the Huronian regions where the quartzites are chiefly found is composed of a matrix of grains of felspar and quartz, together with crystaline fragments of the two minerals (or a quartz-felspar rock) of all sizes from mere grains and chips up to those of pebbles, cobble-stones and boulders. These may be widely scattered in the matrix or crowded closely together, leaving only the interspaces to be filled by the finer debris. The fragments are sometimes quite angular; at others more or less rounded. This is the prevailing rock around lake Temagami, and is also abundant in the whole region drained by the Montreal river. It is also found all the way southward to lake Huron, but in this direction it is often associated with stratified quartzose-diorites and rocks intermediate between the two. The origin of the quartzites appears to be connected with rocks of the above kinds. The fact of their occurrence chiefly in the same regions and in association with them would suggest

Greywacké.

Huronian quartzites.

Origin of the quartzites.

this, but there is also direct proof leading to the same conclusion. The materials forming these greywackés and the stratified quartzose-diorites have been derived from volcanic sources, and coming into contact with water the quartz grains have been by some process separated from the other constituents. In the immediate vicinity of the parent rocks, beds composed more or less completely of these grains are to be found interstratifying other beds formed out of the other constituents. Sometimes these beds are quite thin and shade off vertically into one another, or alternate in great numbers within a limited section. Numerous examples of this arrangement may be seen in the township of Denison and the surrounding country. At greater distances the quartz grains are concentrated in larger volume forming quartzites, while the other ingredients make up the associated felsitic and argillaceous slates with layers of hornblende. The Huronian quartzites all contain grains of felspar in proportions varying from widely scattered particles up to about one-half their volume. Hand specimens of the latter variety bear a strong resemblance both in color and composition to rather fine-grained, reddish and greyish granites or quartz-syenites; but, besides the stratification on the larger scale, the internal structure of the rock is distinctly clastic or fragmental. Examples of these highly felsitic quartzites are to be met with throughout the country north of lake Huron. The coarser varieties are strikingly developed in the highest of the quartzite mountains north-westward of the northern outlet of lake Temagami. In this connection it is worth mentioning that the quartzite beds often found in the vicinity of the phosphate deposits of the Upper Laurentian formation in the county of Ottawa also contain grains of felspar more or less abundantly disseminated, showing that they were probably deposited in a mechanically mixed condition.

Felspar in the quartzites.

The formation of the quartzites being thus apparently connected with the greywackés and quartzose-diorites, they too would seem to partake of the general igneous history of the whole system, which, however, is more obvious in many of its other varieties of rocks. This igneous character is further proved by the large masses or areas of greenstones (diorites or diabases), granites, syenites and other eruptive rocks which are so largely mingled with both the lower and upper portions of the Huronian system in all parts of their distribution, forming indeed one of its characteristic features. The crystalline greenstones occur either as compact areas, wide elongated masses, dykes or thick interstratifying beds, in nearly all the Huronian areas. In many cases the dioritic schists may have been originally massive, but assumed the cleaved structure by pressure when incorporated among stratified masses. The commonest position of the granite and syenite areas is within but towards the borders of the Huronian tracts; but they sometimes occur in the Laurentian country, in their immediate vicinity or at a distance from them in the direction of the longer axis of the Huronian areas.

Igneous character of the system.

An attempt has been made quite lately among some American geologists to restrict the name Huronian to rocks like some of those north of lake Huron, although Sir William Logan and his colleagues in introducing the term originally described it as applying equally to the dark greyish and greenish schists, conglomerates, diorites, etc. The more extended investigations which have since been made in Canada and other parts of the world have confirmed the propriety and convenience of including under this name all the rocks which had been originally described as Huronian.

THE METALLIFEROUS SERIES.

The metalliferous system of Ontario.

The Huronian, as above defined, is the great metalliferous system of Ontario, and indeed of all Canada, and hence its great importance in the economic geology of the country. The whole series is more or less metalliferous, but the various ores are not uniformly distributed, some occurring in one region or in some special stratum, while others may prevail in another section of country or in a different horizon in the series. Besides metallic ores, the Huronian also contains various rocks and non-metallic minerals of value.

IRON.

Occurrence of iron ores.

Iron appears to occur most frequently in the lower or schistose portions of the system. At one place examined by the writer on the Antler river, about 100 miles west-north-west of Port Arthur, there is a large deposit of magnetite of fine quality. In the widest part there are three beds, each about fifty feet in width, separated from each other by only narrow bands of rock, running with the general course of the belt to which they belong. The deposit shows workable quantities of ore at intervals for about three miles, and is traceable for about five miles. No jasper was observed at this locality. Another rich deposit is reported to have been discovered on the Seine river, near its mouth, and one of lower grade ore at a straggling lake at no great distance south-eastward of Wabigoon lake. The iron ores associated with jasper which have been found on the southern part of Hunter's island and near Gunflint lake appear to belong to a continuation of the belt in which the rich deposit of Tower, in the adjoining state of Minnesota, occurs. A belt of fine-grained magnetite in thin layers, alternating with equally thin layers of red jasper, was found and described by the writer in 1869 in the hills on the east side of the Kaministiquia river, just below the place where it is now crossed by the Canadian Pacific railway.* During the same season Mr. Peter McKellar, while assisting the writer to make a topographical and geological survey of lake Nipigon, found a deposit of hematite on its eastern side, near Sturgeon river. A common form of iron ore in the Huronian rocks consists of thin layers of magnetite, or occasionally of hematite, alternating with similar layers of compact or fine-grained grey quartz in the same manner as the jasper just described. These layers vary from one-sixteenth of an inch to an inch and more in thickness, but are usually from about one-eighth to one-half an inch. These ores sometimes occur in large quantities, and although too poor to work (unless some economic process should be discovered for separating the magnetite from the rock), they are nevertheless worth careful examination in the hope of finding the ore in a more concentrated form in some parts. Ores of this character or of a similar class, as far as their economic value is concerned, have been found at the following localities:

Localities.

South-west arm of Red lake, northward of Lake-of-the-woods.
Township of Moss.
Near the height-of-land, south-west of the head of lake St. Joseph, loose.

* See Geological Survey Report for 1869.

Near Little Long lake, west of the north end of Long lake.
Albany river, near the junction of Etow-i-mami river.
The largest of the Slate islands, lake Superior.
Gros Cap, near Michipicoten.
Oka or Pickerel river, west of Michipicoten, reported.
Jackfish bay, reported.
Near Montreal river, lake Superior, reported.
North of Batchawana bay on the Peter Bell location, and in larger quantities at the head of Pancake river.
In one of the south-western bays of Temagami lake.
At a small lake north of the eastern arm of Temagami lake.
Quinze rapids and Opazatika lake, above lake Temiscaming.
At Abittibi lake.

The lean iron ores, composed of thin layers of magnetite interstratifying others of a siliceous rock, in the township of Dalhousie, Lanark county, and adjacent regions are also of this class, and they form another link connecting the rocks of this district with the recognised Huronian. The iron mines of eastern Ontario which were visited by the Commissioners are described in another part of this report.

COPPER AND NICKEL.

Copper is very generally diffused throughout the Huronian rocks, but the principal deposits heretofore worked, those of the Bruce Mines and the Sudbury region, are associated with rocks of the supposed upper division. At the former locality the workings extended for nearly two miles across the Bruce, Wellington and Huron Copper Bay locations. They were carried on chiefly upon two east and west quartz lodes cutting greenstone, the Main lode varying from about three to fifteen feet in thickness, while the other, called the New or Fire lode, is a branch of this. The workings extended to a depth of about seventy fathoms in many parts. The ore was mainly copper pyrites, but a good deal of the purple sulphide was found near the surface. Operations were carried on from 1846 to 1876, and the gross value of the output was ascertained by the writer (through the courtesy of Captain Plummer and others) to be about $3,300,000.

Copper of the Bruce Mines and Sudbury regions.

Copper in the form of the yellow sulphide, associated with pyrrhotite or grey magnetic iron pyrites, is met with in considerable quantities in a number of localities around Sudbury Junction, on the Canadian Pacific railway, and thence along the Sault Ste. Marie branch to the Spanish river. As far as they have been tested these ores also contain sulphide of nickel, often in quantities which should pay to extract. The ore in this region is associated with an obscurely stratified greenstone, and its mode of occurrence is apparently that of large masses and "impregnations," having roughly lenticular forms which resemble "stockwerks" and are rudely conformable with the general stratification of the country rocks. These masses enclose many "boulders" or fragments of all sizes of greenstones and greywacké, which are often finely impregnated with copper and iron pyrites. Most of the ore is of low grade, but in the midst of this considerable bunches of pure copper pyrites occur, while on the other hand some portions consist of almost pure pyrrhotite. The country-rock in this neighborhood everywhere dips at high angles, or is nearly perpendicular, and the ore-masses follow these inclinations. An

Its mode of occurrence in the Sudbury district.

important feature of these masses is their great size, so that although the average percentage of copper and nickel is low (yet sufficient for profitable working), the quantity is so large as to give promise of great productiveness.

First discoveries at Sudbury.

Copper ore was first discovered in this district in 1882, during the construction of the Canadian Pacific railway, at a point on the main line (since called the Murray mine) about three miles north-west of Sudbury Junction. This was soon followed by the finding of the Stobie mine, three and a half miles north, and the Copper-cliff mine an equal distance to the south-west of the junction. At the Stobie mine, which is worked as an open quarry, an ore-mass of the nature above described appears to have a thickness of upwards of a hundred feet and it has been traced south-westward on the strike as a series of lenticular copper-bearing masses, separated by pinched intervals, for a distance of a mile and a-half. The Copper-cliff mine shows a vein-like deposit of mixed pyrites about ten feet wide, containing a higher percentage of copper than the Stobie ore, and also a number of the lens-shaped ore-masses. It has been worked to a depth of over 400 feet on the slope and has yielded a large quantity of good ore. The Evans mine is about a mile south-south-west of the Copper-cliff. Here the ore-bearing mass, as proved by the diamond drill, appears to be nearly 100 feet thick, but it consists largely of pyrrhotite, rich in nickel. In the township of Denison a vein of copper pyrites, with a high percentage of nickel in some parts, has been opened by the Vermilion Mining company on lot 6 of the 4th concession; and on the line between the 5th or Krean lot and the 4th or McConnell lot, both in the 5th concession, and about a mile to the north-north-east of the Vermilion mine, the surface appearances indicate a promising deposit of copper ore similar to those near Sudbury, but no mining has yet been done at this locality.

Stobie, Copper-cliff and Evans mines.

Denison township veins.

The discovery of nickeliferous copper pyrites around Sudbury recalled the fact that similar ore had been found nearly forty years before at the Wallace mine, on the shore of lake Huron near the mouth of the Whitefish river. The deposits at the two localities appear to lie in the same geological horizon, and, in following the general strike of the rocks north-eastward from the Wallace mine, ores of copper have been found near the west end of lake Panache, in the townships of Drury, Denison, Graham, Waters, Snider, McKim and Blezard, on the west side of the Wahnapitæ lake, near the north end of Lady Evelyn lake, on Montreal river, on Blanche river, at a place near the height-of-land not far east of the canoe-route from lake Temiscaming to Abittibi lake, and finally near the south end of lake Mistassini. This will probably prove to be a copper-producing region of vast importance in the future, of which the present discoveries are only the first indications. Copper has been found in many other places in the Huronian rocks of Ontario, and these will be referred to further on in the list of localities of this metal, but the foregoing short descriptions will serve to give an idea of its two principal modes of occurrence in these rocks.

A copper belt extending from lake Huron to lake Mistassini.

GOLD.

Gold promises to become an important product of our Huronian rocks, notwithstanding the fact that only partial success has attended the efforts heretofore made to mine and extract it. If we include the gold-bearing rocks of the Hastings region amongst the Huronian, it becomes doubtful if the precious metal has been found, except in the merest traces, in any other formation. Not long ago the occurrence of gold was unknown in Ontario beyond the traces which had been found in assaying the silver and copper ores of Prince's location, of Michipicoten island, and the vein-stuff of the Bruce mines; but now it has been discovered in so many and such widely separated localities in the province, and in some cases under such promising conditions, that it is highly probable that successful gold mines will be established after more thorough tests have been made.

The importance of recent gold discoveries.

The first discovery of gold in notable quantity was made in 1871 by Mr. Peter McKellar (following up a clue obtained from an Indian) near Jackfish lake, at what is now called the Huronian mine, situated on location H1 in the township of Moss. It here occurs in a true and persistent vein from 6 to 8 feet wide, of which from 2 to 5 feet are quartz, the rest being incorporated schist. The country-rock consists of interbedded talcoid, chloritic, dioritic and a little dolomitic schist, siliceous magnetite and massive diorite, all dipping north-west at angles of 65° to 80°. The vein runs north-eastward, cutting the strata at a small angle and underlying to the north-west side at an inclination of 15° from the perpendicular. Intrusive syenite appears about a mile to the north-east of the mine, and this may have had something to do with the enrichment of the vein. The gold occurs free and as sylvanite (or telluride of gold) associated with galena, iron and copper pyrites and blende, which, with the white quartz, constitute a beautiful looking ore. A 10-stamp mill was erected in 1883 at great expense, on account of the difficulties of transportation, and in 1884 some mining and milling were done. The gold secured is understood to have been equal to $21 to the ton, which was, however, far short of the whole amount contained in the ore. Work was resumed for three or four months in 1885, but, from the want of proper means of transportation to the mine, operations are for the present suspended. Openings have been made and similar ore obtained from a continuation of the same vein, called the Highland mine.

The Huronian gold mine.

Gold was discovered on Lake-of-the-Woods in 1878, or earlier. In the writer's Geological Survey report for 1881, page 15c, it is stated that "in 1879 I was presented by Mr. J. Dewé with a specimen from Hay island, of white quartz containing needle-like crystals of hornblende with a little calc-spar, which showed distinct specks of gold. It was assayed by Mr. Hoffmann, chemist to the survey, and found to contain 37.318 ounces of gold and 1.431 ounces of silver to the ton of 2,000 pounds." During the succeeding four or five years some mining was done at a few places around the northern part of this lake, and in some instances with the prospect of ultimate success, but owing to the impossibility of obtaining titles, on account of the dispute between the Dominion and the Ontario governments as to the ownership of the

Gold locations on Lake-of-the-Woods.

territory, it was impossible to obtain sufficient capital and no thorough test has yet been made to determine the real productiveness or otherwise of any of the mines. Trials have been made at several promising places, such as Sultana island, the Winnipeg Consolidated and the Pine Portage properties, and now that the matter of title is set at rest there is a probability that work will be prosecuted on a sufficient scale to determine the question whether gold is to be found in this region in paying quantities or not. It occurs both free and in combination with sulphides in veins of quartz more or less split up and interrupted, cutting green schists and not far from masses of syenite. These deposits would appear to lie towards the bottom of the series as developed at the Lake-of-the-Woods. Specimens of free gold in quartz have been shown to the writer as having been obtained not far from Taché, on the Canadian Pacific railway.

Partridge lake gold veins.

At Partridge lake, a short distance west of Lac-des-Mille-Lacs, gold was discovered in 1872 by Mr. Archibald McKellar in a large vein of quartz cutting Huronian schist on an islet, and also in large veins of the same material in the strike of this one on either side of the lake. Assays of the quartz from both the islet and the mainland, which were made by Dr. Girdwood, showed from 1 to 1½ ounces of gold to the ton. A number of small nuggets were obtained by breaking up the quartz on the islet by both Mr. McKellar and Mr. W. W. Russell, and shown at the Philadelphia exhibition in 1876. This locality was visited by the writer last summer on behalf of the Commission, and samples of the quartz and photographs of the outcrop of the vein were obtained.

Victoria Cape location.

In 1875 Mr. Donald McKellar found small nuggets of gold in a vein of quartz cutting reddish granite at Victoria cape on the western side of Jackfish bay, on the north shore of lake Superior. Another vein of quartz 1½ to 3½ feet thick, holding iron pyrites, galena and blende and cutting the granite in close proximity to slaty diorite at this locality, yielded on assay $27 worth of gold per ton.

Various other locations.

The Commissioners were credibly informed that gold associated with iron pyrites had been discovered in quartz veins between Goulais and Batchawana bays, at the east end of lake Superior, and also on a small island near the shore of lake Huron, north of Lacloche island. Traces of the metal have been found by assay in the laboratory of the Geological Survey in quartz from veins near the north end of Temagami lake, and from Cross lake to the south of this sheet of water.

Vermilion Mine location.

At the Vermilion gold mine on lot 6 in the 4th concession of the township of Denison, in the Sudbury district, coarse free gold was found at and near the surface in a vein of light grey granular quartz about two feet thick, running north-eastward, and cutting grey or ash-colored greywacké, bearing a close resemblance to the "whin-rock" of the Nova Scotia gold districts. The gold was so thickly disseminated in one part of the vein as to hold together fragments of the quartz after they had been fractured by the hammer. On the same lot a ridge of greyish diorite rises a short distance to the south-east of this vein, and a few specks of gold were seen by the writer in the midst of iron-stained spots on the weathered surface of this rock. Visible gold is said to have been found also in a quartz vein at the east end of the Indian reserve on the south side of the mouth of the Spanish river.

Gold was discovered in the summer of 1888 on a point on the southern shore of lake Wahnapitæ, between the two deep bays on that side of the lake. It occurs, as far as could be observed by the writer, who visited the place in September, in several narrow veins of white quartz cutting a highly felspathic reddish quartzite, resembling fine grained granite, but distinctly clastic or fragmental in origin. No single vein or group of small veins could be traced far, but where any of them gave out others were observed to commence not far off. They are "bound" veins, or adhere closely to the wall-rock, or are, as it were, fused into it. One of these little veins shows a good deal of mispickel and some iron pyrites in crystals along one side of it. An assay of a sample from this vein, made by Mr. Hoffmann, chemist of the Geological Survey, yielded at the rate of 5.425 ounces of gold and 0.233 of an ounce of silver to the ton of 2,000 pounds, while the quartz from another of these veins showed neither gold nor silver. The visible gold of these veins occurs as specks and small nuggets in the quartz. It is also said to have been detected in the wall-rock apart from the veins, and, if so, this circumstance may prove to be of great importance, for if it should be found to pay at all to treat the whole mass, great profits might be realised by doing so on a large scale. The rock at this locality resembles that of the famous Treadwell mine on Douglas island in Alaska, both in composition and in the manner of the occurrence of the small adherent quartz veins, but the latter rock is almost white and is dotted with small particles of iron pyrites. A microscopic examination of the ore of the Treadwell mine, which has just been made by Mr. F. D. Adams of the Geological Survey, shows it to consist of a granite-like rock in which the clastic character may be due to a process of crushing after its solidification. Although the rock of the Treadwell mine yields only about $5 worth of gold to the ton, immense profits are made by stamping and treating it in large quantities, and it is in this direction rather than in the great richness of small quantities that we must look for profitable gold mines in our Huronian rocks. It may be here remarked that gold is said to have been profitably extracted from two mines in Huronian rocks on the south side of lake Superior.

Gold on lake Wahnapitæ.

Comparison with Treadwell mine, Alaska.

In the early part of August, 1866, gold was discovered by a man named Powell and a Dutch miner on the eastern part of lot 18, range 5, in the township of Madoc, belonging to Mr. J. Richardson, who, however, did not recognise it as the precious metal till informed of the fact by the late Mr. H. G. Vennor of the Geological Survey, who was then working in the neighborhood. Mr. Vennor in his report for that year, addressed to Sir William Logan, described the gold as occurring in "a series of crevices or openings in a gold-bearing bed, formed of chloritic and epidotic gneiss (or schist) holding patches of dolomite and calcspar, the openings being nothing more than such as are so often met with in the dolomites and calc-schists of this region." The gold was found along with particles of black carbonaceous matter in a

Gold in the Hastings region

Richardson mine.

brown ferruginous earth filling the longitudinal crevices, paralled to the bedding, one of which had been struck at a depth of 4 and another at 15 feet from the surface at the time of Mr. Vennor's visit. Numerous small nuggets were also found enclosed in the adjacent dolomite and calcspar. The strata here dip nearly due north at an angle of 45°, and the gold-bearing bed is "overlaid by a siliceous ferruginous dolomite and underlaid by a band resembling an impure steatite." Its geological position is not far above the iron-bearing belt of that region. The Richardson mine has been worked at different times since the above date, and a good deal of gold extracted from it.

Other localities of gold.

This discovery was followed by many others of the precious metal which have been made at different times in the townships of Marmora, Madoc, Elzevir, Kaladar, Lake and Tudor, and there is now a probability of gold-mining becoming an established industry in this region. One of the most notable of the attempts at gold mining in the district is that at the Gatling (since called the Canada Consolidated) mine in the township of Marmora.

Gatling mine.

The gold here occurs in veins of quartz containing much mispickel and cutting a crushed syenite or a mixture of schist and syenite, close to a large area of the latter rock. Assays of twelve different samples of the ores of this mine gave an average of 1.9107 ounces or $39.47 to the ton of 2,000 pounds. In spite of this richness, the difficulty of separating the gold from the sulphide of arsenic is so great that only partial success has attended the working of the mine, after the expenditure of a large sum of money in buildings, machinery, working the mine and experimenting.

Gladstone and Feigle mine.

A considerable quantity of gold has been extracted from the Gladstone and Feigle mine, situated on the continuation of the same set of veins as the Canada Consolidated, at a distance of two or three miles to the northward of it.

Dean and Williams mine.

Another mine called the Dean and Williams on lot 8, range 9 of Marmora, about a mile and a-half southward of the Canada Consolidated, was worked for a time with some success.

Guinard mine, Kaladar.

At present it is reported that from six to eight dollars worth of gold per ton are being extracted at the Guinard mine, in Kaladar, from a set of small quartz veins cutting a rock which is described as a conglomerate with quartz pebbles in a matrix of micaceous schist.

SILVER.

Silver in Huronian rocks.

Some veins have been discovered in the Huronian rocks containing silver in promising quantities—sufficiently so at any rate to warrant the search for more. The vein at the Huronian mine is a case in point, but the silver there was pratically overlooked in the efforts to extract the gold.

3A mine.

The 3A mine on the north shore of Thunder bay was opened on a vein of quartz and bitter-spar from 1½ to 2½ feet wide and running about east-north-east, or parallel to the strata, which, according to Mr. Peter McKellar, here "consist of thick beds of diorite and fine grained greenish-grey slates, some of which are chloritic, talcose, calcareous and ferruginous, with some serpentine alongside of and in the vein." Dark greyish red syenite is met with a short distance to the south. The silver occurs both native and in combination with sulphur and nickel, and it is associated with iron and copper pyrites, galena and blende. There is also a small proportion of gold along with the silver.

This vein was discovered in 1870. Active operations were begun early in the spring of 1873, and after having been worked to a depth of about 150 feet and yielding several thousand dollars' worth of silver in the form of bunches of very rich ore, work was suspended in the spring of 1874. Two veins were discovered in 1872 within a mile of Heron bay and close to the Pic river, which, judging from the ore brought to the shore of lake Superior and examined by the writer, bear a strong resemblance to that of the Huronian mine. The width of each is given by Mr. Peter McKellar as ranging from a foot to four feet at the surface. They lie in a large Huronian area and are described as cutting talcoid and chloritic schists, while a boss of intrusive granite rises beside the Pic river at no great distance to the eastward, giving an additional point of resemblance to the surroundings of the Huronian mine. One of the veins runs east-north-east with the stratification, while the other strikes nearly north and south. At a depth of 40 feet the latter had opened out to 5 or 6 feet in width. The gangue in both veins consists of bitter-spar and quartz, and contains galena, blende, iron and copper pyrites, together with gold and silver, ranging from traces up to about $70 worth of each to the ton, according to assays made by Mr. McDermid, who was assayer at the Silver Islet mine at the time the above work was done. Heron bay.

LEAD.

The Victoria argentiferous galena vein, situated near Garden river and about eight miles north from its mouth, occurs in Huronian rocks. It runs about north-north-west, parallel to the western side of an extensive mass of very fine-grained reddish-grey granite or quartz-felspar rock, from which it is separated by a few feet of glossy green schist and tough green trappean rock, some of the latter approaching the character of an amygdaloid. Work was commenced at the Victoria mine in 1875, and at the time of the writer's visit in 1876 two shafts had been sunk, each to a depth of 15 feet, in the midst of a belt 36 feet thick, of glossy-surfaced green schist, cleaving in all directions and containing galena in strings, grains and small bunches. One of the shafts followed a vein of solid galena, mixed with considerable dark blende and a little copper and iron pyrites, from 8 to 19 inches thick, and the other a similar vein 10 inches thick, but containing a mixture of quartz. This lead-bearing belt of schist is succeeded on the west by siliceous felsites and dark green and rather coarsely crystalline hornblende-rock, which is again followed by fine-grained light reddish or pinkish-grey granite. This belt of veins was afterwards worked to a considerable depth, and a large quantity of galena taken out and exported. The proportion of silver varied from a few ounces up to 168 to the ton of 2,000 pounds, most of the ore being tolerably rich. The Cascade mine, a short distance to the northward of the Victoria, is said to be on the same belt and to resemble the latter in most respects. Argentiferous galena has also been found in Huronian rocks in the Sudbury district, a short distance south of Straight lake, and in the north-western part of the township of Creighton, and again near the north end of Lady Evelyn lake, which lies between Temagami lake and the Montreal river. Victoria mine, Garden river. Cascade mine. Sudbury district.

The Victoria and Cascade mines are the only places at which any lead mining has been done in Huronian rocks, unless the galena veins of Tudor, Limerick, etc. in the Hastings-Lanark region should prove to be situated in rocks of this age.

Other galena veins.

ZINC.

Zinc, in the form of blende (or sulphide of the metal), was discovered in 1881 in large quantities at a place which has been named the Zenith mine, situated on the White Sand river, about ten miles northward of the shore of lake Superior opposite Wilson's island, eastward of Nipigon bay. The ore is black and crystalline, and is described as occurring principally in two large veins or lenticular masses in a hornblende rock or diorite of Huronian age. The ore could be mined with great facility, and some 400 or 500 tons are said to have been already excavated, but it cannot be brought to market for want of a road. A specimen analysed by Mr. Hoffmann, chemist to the Geological Survey, gave 54¼ per cent. of metallic zinc. This is the only locality at which zinc ore has been discovered in large masses in the Huronian system in Canada, and it is therefore interesting as an indication that the metal may be found in paying quantities in these rocks in other places. Blende in large crystals occurs in a vein of coarse calcspar about eight feet wide at Blende lake, about 1½ miles north-north-west of the head of Thunder bay. The south wall of the vein, which runs east and west, consists of dioritic schist of Huronian age, while the north wall is formed by ferruginous and siliceous clay-slates belonging to the Animikie series. The occurrence of blende with the galena of the Victoria mine has been already referred to.

Zinc at Zenith mine.

Blende lake, Thunder bay.

ANTIMONY.

In the report of the writer for 1876, page 211 (Geological Survey Report) reference is made to a reputed discovery of sulphide of antimony in a vein of white quartz cutting felspathic grey quartzite (Huronian) about one mile west of Fairy lake, near Echo lake. Mr. Joseph Cozens, of Sault Ste. Marie, states that he has discovered a vein eight inches wide, rich in this ore, among the Huronian rocks on Garden river.

Antimony near Echo lake and Garden river.

OTHER METALS AND MINERALS.

The above descriptions will serve to show that the Huronian rocks contain ores of the various metals referred to in economic quantities. The occurrence in them of nickel, arsenic and tellurium has been incidentally mentioned. In addition to the metals already alluded to, platinum, tin, molybdenum, bismuth and cobalt have also been found among these rocks, but our space will not admit of a fuller description of them than that contained in the list further on. The number of discoveries of valuable ores already made must be regarded as very encouraging, considering how little knowledge we possess as to the geological relations and modes of occurrence of the metals in the Huronian system and the comparatively small amount of *bona fide* and intelligent exploration which has yet been done, and leads to the belief that many districts situated on this extensive system will prove rich in metallic ores.

Other metals found in Huronian rocks.

Among the rocks and non-metallic minerals of economic value to be met with in the Huronian system, the following may be mentioned: fine granites and syenites for buildings, monuments and ornamental purposes; sandstones, quartzites and greywackés for construction; flagstones, roofing slates, serpentine and dolomitic marbles, ornamental argillites, jasper conglomerate; white quartzite for glass-making; asbestos (chrysotile), graphite, actinolite and barytes.

Useful rocks and non-metallic minerals.

THE CAMBRIAN SYSTEM.

This system was so called by the late Professor Sedgwick from Cambria, the ancient name for Wales. It is the oldest one in which the remains of organic life have been found, being the first above the Archæan rocks. Besides ordinary sedimentary deposits, such as limestones, sandstones, shales, etc., it comprises, in some regions, a large proportion of igneous and other non-fossiliferous rocks, the whole thickness of the system amounting to many thousands of feet. The fossils, when present, consist only of marine invertebrate animals, among which trilobites are conspicuous, both as to size and the number of species. Remains of marine plants are also found in these rocks. The system is well developed in Bohemia, Wales, Newfoundland, New Brunswick and Ontario, as well as in some parts of the United States.

Rocks composing the Cambrian system.

ANIMIKIE FORMATION.

The name Animikie is derived from the Outchipwai word for thunder, and is appropriate inasmuch as these rocks are most largely developed around Thunder bay. Although no fossils have yet been found in the Animikie formation, it consists largely of undisturbed and unaltered sedimentary rocks and is classified with the Cambrian. It is the first formation above the Archæan, and rests almost horizontally upon the denuded edges of the upturned Huronian and Laurentian strata of the region. Its thickness has not been clearly ascertained, but it probably amounts to 2,500 or 3,000 feet or more.*

Derivation of Animikie.

The Animikie strata in ascending order consist of: (1) Greenish arenaceous conglomerate with pebbles of quartz, jasper and slate—seen on the north shore of Thunder bay. (2) Thinly bedded cherts, mostly of dark colors with argillaceous and dolomitic beds—seen at the head of Thunder bay and along the northern boundary of the formation in the township of McIntyre. (3) Black and grey argillites and flaggy black shales with sandstones and ferruginous dolomitic bands and arenaceous beds, often rich in magnetic iron,

Nature of the rocks.

* The Silver Islet mine was sunk through 1,230 feet of these rocks lying almost horizontally below the level of lake Superior, and on the main shore opposite to the mine they rise about 800 or 900 feet above the lake, in addition to a trap overflow of 400 or 500 feet, which however may belong to the next higher formation; so that we have here an actual section of some 2,000 feet. Mr. E. D. Ingall says (Report of Geological Survey for 1887, page 25 H): "If we assume the average dip to be in a south-south-east direction and measure the width of the outcrop of the formation from Grand Portage island where it passes under the overlying Keweenawan rocks north-north-west to near Woodside's vein in the Silver mountain area where the Archæan appears from below them, which we find to be some 25 miles, we get a thickness for these rocks of over 12,000 feet." But on the same page he says, "As has been already mentioned, the formation lies nearly flat, and it is very difficult to decide whether it really has any general dip or not."

together with layers and intrusive masses of trap (diabase). This is by far the thickest division, constituting in fact the bulk of the formation. Lenticular and spheroidal concretions of various sizes, called also bombs, boulders and kettles, are common throughout the black shales of this division.

Geographical distribution of the formation.

Geographically, the Animikie formation in Ontario occupies a triangular area, of which the base, 60 miles in length, extends from the mouth of Pigeon river westward along the international boundary to Gunflint lake, while one of the other sides is formed by the shore of lake Superior from Pigeon river to Goose point on Thunder bay, 40 miles in length, and the third side by a line joining this point and Gunflint lake, about 80 miles in length. The lower portions of the formation extend over the comparatively level ground within this area to the northward of the Kaministiquia and Whitefish rivers, while the higher measures, lying almost horizontally, occupy the mountainous country stretching from these streams southeastward to the shore of lake Superior. The summits of most of the hills in this district are capped with thick and nearly horizontal beds of trap, giving them a flat or table-topped appearance. If all these isolated areas of trap were represented on a geological map of the district it would have a "spotty" appearance, the total extent of the trap being much less than that of the underlying shales which occupy the hill-slopes and the bottoms of the valleys between them. A good idea of the geology of this district may be formed if we suppose the crowning overflow to have been once continuous, but that afterwards extensive erosion of both the trap and the underlying shales took place, leaving only the detatched portions or islands we now see. There is evidence of some faulting in various parts of the district which may account for the difference in level of some of these trappean cappings. The crowning overflows are, however, not the only beds of trap which exist in the formation, as was pointed out many years ago by Sir William Logan and the writer, and more recently by Mr. Ingall. Examples of this may be seen along the Pacific railway track east of Port Arthur, on the islands and points of the north side of Thunder bay, at the mouth of Current river, and between the Duncan mine and Port Arthur, as well as within the limits of that town itself. One of these beds of trap forms a conspicuous escarpment with a long slope to the southward on mining-lot L in the township of McIntyre, which may be seen from Thunder bay. The nearly horizontal layers or masses of trap in these rocks may not have been surface-flows in all cases. Some of them appear as if they might have been injected under pressure between the bedding of the shales and other rocks, and in such cases they do not seem to extend very far.

Table-topped hills.

Several beds of trap.

Other localities of Animikie rocks.

The Animikie rocks are found near the water's edge along the south-east side of Thunder bay, as far as Thunder cape and around its southern side, including the islands to a point east of Silver islet; also in places along the main shore and on some of the islands about Rossport, east of Nipigon bay. They likewise form Pie island and the chain of islands extending thence near the coast to Pigeon river. The Animikie rocks do not appear to extend far into Minnesota, being replaced by higher strata south of the boundary.

They are said, however, to recur on the south side of lake Superior in the northern parts of Wisconsin and Michigan, but in a disturbed condition and otherwise differing from the Canadian type.

THE SILVER-BEARING ROCKS.

The Animikie rocks are of great importance as being the silver-bearing formation of Ontario. Nearly all the veins cutting these rocks bear a strong resemblance to one another in the nature of the gangue which fills them, so much so that these veinstones may be said to form one of the characteristics of the formation. They are generally open or drusy, brecciated and very crystalline, consisting in most cases of white quartz and calcspar mixed with fragments of the wall-rock and a smaller proportion of green and purple fluorspar; but in some instances, as at McKellar's island and in most of the veins on the islands and mainland between Thunder bay and Pigeon point, white barytes forms one of the principal constituents. A part of the crystalline quartz is almost invariably amethystine, and this color has also been imparted to the quartz crystals of the veins in the older rocks of the neighborhood, once covered by the Animikie, which is not the case in similar veins at a distance, showing that the character of veins may be influenced by the rocks above. The veins of the Animikie formation are apt to be gathered into solid bodies where they traverse the harder beds and the trap layers, and to become scattered into a number of strings or reduced to mere faults, or to be pinched out altogether in passing through the softer strata, such as the black shales.

The silver-bearing formation of Ontario.

Variations in veins.

The silver occurs native in grains, threads and small branching forms, and as argentite in leaves and small masses, but occasionally in large crystalline lumps, as at the Rabbit Mountain mine. At Silver islet heavier native silver and two new silver compounds, Huntilite and Animikite, were found. The associated sulphides in nearly all the veins are blende, galena, copper and iron pyrites, and their relative proneness to carry silver has been found by many assays to be in the order in which they are here mentioned. In the same way they have also been found to be richer in silver in proportion as they are more closely associated with the visible silver itself, and to contain very little when remote from the rich ore of the vein. No law governing the conditions or mode of occurrence of the silver in the veins has yet been discovered, and its apparently sporadic distribution in them makes it necessary to prospect extensively underground in each case before a vein can be pronounced valuable or otherwise.

Forms of silver.

Uncertain distribution of silver.

The veins in the Animikie rocks run principally in two general directions, one about north-east or east-north-east to east, and the other about north-north-west. The veins of the Beck or Silver Harbor, the Thunder Bay and Duncan mines and the majority of those in the townships of McIntyre and Neebing and in the Rabbit mountain and Silver mountain districts belong to the former, while those of the Beaver and Silver islet mines and the numerous veins traversing the islands and the mining locations on the mainland between Thunder bay and Pigeon point belong to the north-north-west group.

Courses of veins

The first discovery of silver of any consequence on lake Superior was made by Mr. Peter McKellar in the autumn of 1866, at what afterwards

Thunder Bay mine.

became the Thunder Bay mine. The metal was here found in the form of grains and threads of native silver thickly disseminated in a vein of light grey granular quartz from 1 to 3 feet wide, running north 34° east and cutting dark shale and argillite, interstratified with impure ferruginous dolomite and overlaid by a bed of trap. This vein was worked in the summers of 1869 and 1870, in each of which years it was visited by the writer. Notwithstanding its great richness at the surface it failed to produce any large amount of silver, and work was discontinued after reaching a depth of only 70 feet in each of the two shafts which were sunk. Operations were resumed in 1875 on a parallel reticulated vein in argillite, having a breadth of 6 to 12 feet in all, which was struck by cross-cutting underground at a distance of about 20 feet south of the silver-bearing vein. This opening was inspected by the writer in 1875, but no silver could be detected. Further exploration along the course of these veins may give better results.

Shuniah or Duncan mine.

The vein of the Shuniah, afterwards called the Duncan mine, was discovered to be silver-bearing by Mr. John McKellar and Mr. George A. McVicar in May, 1867, or about a year after the discovery of the Thunder Bay mine. The vein is very large, being 20 to 30 feet in width, runs east and west and consists of quartz and calcspar. The Duncan mining company showed great courage and perseverance in working this property, and only abandoned it after spending about half a million dollars and sinking to a depth of 800 feet, with galleries at different levels. Only about $20,000 worth of silver in all was obtained. The Animikie rocks were here found to extend to about half the above depth, and below this were Huronian schists with syenite on or near the north wall of the vein. A considerable area of trap occurs on the surface to the south of the vein.

Silver Harbor or Beck mine.

The vein at the Silver Harbor or Beck mine was discovered by Ambrose Cyrette in 1870, and was worked in 1871 and 1872. It cuts the shales and chert beds of the lower part of the formation, which are here overlaid by a bed of trap. It runs east-north-east, and is about 5 feet in thickness. The gangue is of a brecciated character, and consists of white quartz and calcspar with a little fluorspar and amethyst. Some native silver was obtained in association with galena, blende and iron pyrites, but the quantity does not appear to have been remunerative.

Discovery of Silver Islet mine.

The Montreal mining company having employed Mr. Thomas Macfarlane, a well known geologist, to survey and prospect their locations on lake Superior in 1868, his assistant, Mr. Gerald Brown, while triangulating the shore of Wood's location, sent one of his men, John Morgan, to plant a picket on a rock in the water, afterwards called Silver islet. While doing this Morgan found the silver-bearing part of this now famous vein,* and brought a specimen from it to Mr. Brown. Mr. Macfarlane took out some $1,500 worth of silver with one or two blasts at the surface and sent it to Montreal. The vein was further prospected to a small extent by the Montreal company the next year, but in 1870 this property, along with all the other locations of the company,

* In the Geology of Canada, page 707, Sir William Logan describes the Silver islet vein where it crosses Burnt island as "a very prominent lode holding galena and green carbonate of copper."

passed into the hands of the Ontario Mineral Lands company, by whom it was worked till the beginning of 1884, when a depth of 1,230 feet had been reached and silver to the value of $3,250,000 had been extracted. The vein in the part worked would average about 8 to 10 feet in thickness, although in some places it measured from 20 to 30 feet. Its course is north 32° to 35° west, and it intersects a dyke of trap (diabase) running east-north-east, which cuts the dark shales and other nearly horizontal strata of the Animikie formation. The silver was found only in and near the trap, which no doubt had something to do with its concentration at this place. Its deposition also appears to have been influenced by graphite, which was present in the richest parts of the vein. Hydrocarbon gas and water holding chlorides of sodium, calcium and magnesium were struck in the deeper workings of the mine. Graphite and inflammable gas have since been met with in other silver mines in the district. The Silver islet vein is easily traceable across Burnt island and upon the main shore opposite, but it was not found sufficiently rich in silver to be worth working except at the islet.

Description of the vein.

The discovery of silver-bearing veins in the Rabbit and Silver mountain districts about 1882 was due to an Indian named Tchiatang who had worked with the writer in the Thunder Bay district and around lake Nipigon in 1869. He is a native of unusual intelligence, and after observing our operations and making many enquiries about veins, etc., he developed a strong ambition to prospect for minerals. While exploring in the neighborhood of Rabbit mountain he discovered the vein which was afterwards worked there, but on account of an Indian superstition* he would not personally point it out to a white man. Mr. Oliver Daunais had married his daughter, and he got over the difficulty by taking him nearly to the place and explaining where he would find the vein. He afterwards revealed in the same way other discoveries which he made in this district, and these have led to all the present developments, so that the latter are indirectly due to the operations of the Geological Survey. Several mines in this part of the Animikie area have been successfully worked and have yielded large amounts of silver, the most conspicuous examples being the Rabbit Mountain, the Beaver, the Badger and the West End Silver Mountain, but as these are all described along with the other mines of this section in another part of the present report they require no further notice here.

Discovery of Rabbit and Silver Mountain veins.

Other silver mines.

Geologists and miners have speculated a good deal as to the source of the silver and the conditions which have influenced its deposition in certain parts of the veins cutting the strata of this formation. Some, like Mr. Peter McKellar, suppose it has come up from the Huronian system below, and they instance the 3 A mine as an example of the occurrence of silver in a vein traversing these rocks. Others suppose the upper parts of the formation to be the parent rocks. The silver appears to be most abundant in those portions of the veins which lie immediately under existing trap beds or their former extensions, or, as at Silver islet, in a dyke of the same rock, and this would favor the idea that the metal has been in some way

Origin of the silver.

* Indians believe that if they show a discovery of valuable minerals to a white man they are sure to die within one year.

derived from the trap. If this be the correct view we may look for silver in the veins under trap beds in the lower as well as in the upper parts of the formation. Indeed this appears to have been demonstrated in the case of the rich ore at the Thunder Bay mine and in veins recently discovered in the Whitefish lake region, both of which are in the lower portions. In these instances the characters of both gangue and ore are the same as in veins higher up in the formation, as if the nature of their contents depended more on the overlying trap layers than on the immediately enclosing rocks. The enrichment of the 3 A vein may have been due to the infiltration downward of the silver from Animikie rocks, which once covered this spot as they still do the immediate vicinity, but which have since been carried away by denudation.

NIPIGON FORMATION.

General characteristics.

A voluminous set of rocks to which this name was given by the writer in 1872 is found in the lake Superior region, resting apparently unconformably upon the Animikie formation. It is characterised by reddish marls, sandstones and conglomerates, together with a large proportion of variously colored trappean beds and masses, a considerable part of which is amygdaloidal. It is largely developed on Black and Nipigon bays, around lake Nipigon and on Michipicoten island and the promontory of Mamainse (more correctly Namainse). The Keweenawan, or native copper-bearing rocks of Keweenaw point, on the south side of lake Superior, and of Isle Royale, appear to be of the same age. On the north shore native copper has not yet been discovered in such large quantities as on the south side. Its relative scarcity may be due to the less disturbed condition of the rocks on the former. On Isle Royale and Michipicoten island, where the dips are highest, the metal has been met with in the most promising quantities. In the Keweenaw point region the metallic copper has been found as large masses in veins cutting trappean beds, and at the Calumet, Hecla and Tamarac mines in the form of grains and straggling masses of all sizes, filling the interstices of a reddish arenaceous mixed breccia and conglomerate bed, in which many of the fragments consist of red quartziferous porphyry.

Keweenawan rocks.

Native copper on lake Superior.

On the north shore.

Native copper has been met with, principally in veins, on St. Ignace, Simpson's and Battle islands. On Michipicoten island it occurs as grains in an arenaceous ash-bed, while at Namainse it is found in the form of straggling masses and thin sheets in calcspar veins. But the rocks which have been classified as the Nipigon formation are not alone copper-bearing, but contain other metals as well, and they differ in this respect in the several regions in which they are developed. For example, the lead-bearing character of the red marls and associated rocks to the north-west and north of Black bay is their most important feature from an economic point of view. No metallic ores of any consequence have yet been discovered in the rocks of this formation around lake Nipigon.

Lead in Nipigon rocks.

On the west side of Little Pic river certain trappean strata, distinctly bedded and dipping regularly to the south-westward at an angle of about 12°, may belong to this formation, if not to the Animikie. One bed is amygdaloidal, but all the others are crystalline and of different shades of

Little Pic river.

color. These rocks may have a thickness of 600' feet, and they occupy a space of about eight miles along the river and fourteen on the lake front. On the west side of the mouth of the river horizontal beds of siliceous magnetic iron ore occur in these rocks. The united thickness of three of them appears to be about ninety feet. The percentage of metallic iron contained in the ore from this locality was found to be as follows: by Dr. Hayes, of Boston, 36; by Dr. Girdwood, of Montreal, 46; and by Dr. T. S. Hunt, 37, chiefly as silicate.

Siliceous magnetite.

In the region extending from Thunder bay to Nipigon bay, and thence northward, the strata of this formation consist of the following rocks in ascending order: white grits (seen in the cliff along the south-east side of Thunder bay), red and white sandstone with conglomerate beds, the pebbles being mostly of jasper in a sandy matrix of different colors, compact argillaceous limestones, shales, sandstones and red indurated marls, red and white sandstones with green spots, and red and white conglomerates interstratified with trap layers. These are covered by an enormous amount of trappean overflow crowning the formation, and amounting to from 6,000 to 10,000 feet in thickness. The lower portions of the overflow are usually massive and crystalline, but it becomes more amygdaloidal towards the top. Much of it consists of columnar basalt, but varieties of pitchstone also occur. This volcanic portion of the formation is well developed on St. Ignace and Simpson's islands, and also about point Porphyry. The amygdaloids contain agates and quartz, prehnite, epidote, specular iron, various zeolites and native copper. Concentric wrinkles on the surface of a trap bed on St. Ignace island show that a flowing movement took place in a north-easterly direction, while on the east side of lake Superior similar wrinkles mark an easterly motion of the molten trap. Flow-wrinkles of the same kind occur at the Calumet mine on the south shore. Dykes of greenstone are very numerous in the upper part of this formation. They run parallel to one another in two principal directions, and they all have a transverse columnar structure. There are also many dykes of porphyry in these rocks.

Rocks of Nipigon formation.

Trappean overflow.

Concentric flow-wrinkles.

Dykes.

The crowning overflow of columnar trap on Thunder cape, Pie island, McKay's mountain and all the other hills between the Kaministiquia and Pigeon rivers looks quite like that resting on the red marls of Nipigon bay, Black Sturgeon river and lake Nipigon, and no geological distinction between them has yet been pointed out.

Trap of Thunder cape, McKay's mountain, etc.

The lower members of the Nipigon formation occupy most of the peninsula between Thunder and Black bays, and extend northward up the valley of the Black Sturgeon river to lake Nipigon and around this sheet of water. The grey sandstones, indurated red marls and variously colored compact limestones are well developed in the township of Dorion and to the westward of it. In the north-western part of this township a fault running east-north-eastward has let down the red marls on its south-east side against Laurentian gniess on the north-west. A well defined brecciated vein of quartz, calcspar and barytes, holding galena and a little copper pyrites and blende, has been traced for more than two miles on and near the junction of the two rocks. It varies from fifteen to twenty-five feet in width and is very

Geographical distribution.

Lead veins in Dorion township.

conspicuously exposed, owing to its standing the denuding agencies better than the enclosing marl. The locations of Mr. C. J. Johnson and J. S. Turnbull are situated on this vein. The Malhiot vein runs nearly parallel to the last, at a distance of about three miles to the south of it. It cuts

Malhiot vein.

crystalline trap rocks resting on nearly horizontal compact limestones and grey sandstones. Its width varies from six to eight feet, and towards the western part of its outcrop it is well charged with galena in a gangue of of calcspar, quartz and barytes. A large lead-bearing vein, also parallel to the two above described, crosses the lake at the head of Wolf river. It cuts Laurentian gniess in the valley under the marls and trap which form the cliffs on either side of the lake. Another lead-bearing vein, which has been styled the Ogama, is reported in Dorion about five miles south-west of the Malhiot.

Lead veins near Black bay.

In May, 1865, Messrs. Peter and Donald McKellar discovered an important vein of galena cutting the indurated red marl of this formation at a place about three miles west of Black bay, on what is now called lot C in the township of McTavish. The property has been successively named the North Shore, Lead Hills and Enterprise mine. The vein runs about north 60° east, and the red marl is here associated with grey sandstone; but red granite, which is largely developed in this region, rises as a low bluff about 300 yards to the north of it and was encountered at a moderate depth in working the vein. The gangue is quartz, calcspar and barytes, and the total width of the vein is from six to eight feet, of which from three to four feet consisted for some distance of solid galena with a little copper pyrites and vein matter. The mine was worked for one year, and a considerable quantity of rich ore was shipped to the United States. According to assays made by Prof. Chapman it contained an average of $17 worth of gold and $2 worth of silver to the ton.* A vein carrying galena has been discovered in the Nipigon rocks between Pearl River station on the Pacific railway and the shore of Black bay. More than forty years ago Sir William Logan found galena in the rock of Granite island in Black bay.

Lake Nipigon.

Crystalline columnar traps similar to those which overlie the red marls, etc., of Nipigon bay are largely developed around lake Nipigon, where they lie for the most part horizontally, and form the prominent bluffs and islands which give rise to the picturesque scenery of the lake. Compact limestones and grey sandstones are found under these traps in several places, and in the hill just behind Nipigon House, on the west side, a stratified red felspar rock, studded with grains of vitreous quartz and having a probable thickness of about a thousand feet, dips north-north-west at angles varying from 40 to 60°. A massive rock, but of a similar lithological character, occupies the lake shore from Nipigon House to English bay, a distance of three miles.† This rock resembles the red quartziferous porphyry, which forms so many of the pebbles in the native copper-bearing conglomerate of the Calumet mine.

Michipicoten island.

Michipicoten island may be described in a general way as consisting mainly of trappean beds, dipping about south-by-east at an angle of 30°.

*Dr. Bell in Geological Survey Report for 1869, p. 359, and for 1872, p. 108.
†Report of Geological Survey for 1871, p. 103.

Along its northern shore they are mostly amygdaloidal, and are here associated with trap-conglomerates and red sandstones and shales, passing below the trappean mass to the south. Approaching the southern side the ordinary varieties are overlaid by compact reddish trap, sometimes rendered porphyritic by crystals of red felspar and white quartz. Along the south side the trap becomes black and has a resinous fracture. In this part of the island there are some amygdaloidal beds, with fine agates suitable for ornamental stones. The whole thickness of the strata on Michipicoten island probably amounts to about 12,000 feet. On the western end of the island a mine has been opened on a bed carrying native copper, which is so fully described in the evidence of Mr. Joseph Cozens that no further notice of it is required.

Mine of native copper.

The promontory of Namainse on the east side of lake Superior is occupied by rocks of the Nipigon formation. They comprise a variety of amygdaloids, volcanic tufas, felsites, cherts, sandstones, coarse conglomerates and crystalline traps. The dip is to the westward, or into lake Superior, at an average angle of 45°. At a moderate calculation the thickness of the strata on this promontory would amount to 22,400 feet. The conglomerate bands form one of the most striking features of these rocks, both on account of their coarseness and the thickness of the beds. Five of them occurring among the amygdaloids, tufas and crystalline traps south of Point aux Mines measure respectively about 260, 85, 70, 80 and 450 feet. Most of the enclosed masses are well rounded and smooth, and from the large size of many of them these beds may be properly called boulder conglomerates. They consist principally of dull reddish granite and greenish and greyish Huronian schists.* Nipigon rocks, like some of those on Namainse, are met with at Batchawana bay, and a small patch of them at Gros Cap at the outlet of lake Superior.

Promontory of Namainse.

Boulder-conglomerates.

Batchawana bay.

On the east side of Hudson bay and the islands lying off that coast volcanic and sedimentary rocks are largely developed. They comprise reddish conglomerates and sandstones, lead-bearing limestones, chert-breccias, black shales, grey quartzites, dark argillites, porphyries, crystalline traps, amygdaloids, tufas, etc. The upper parts of this series may correspond to the Nipigon formation and the lower to the Animikie.†

Nipigon and Animikie formations on Hudson bay.

POTSDAM FORMATION.

In Ontario this formation consists almost entirely of hard grey and sometimes reddish sandstones. It derives its name from the town of Potsdam in the north-eastern part of the state of New York, and was called the Potsdam sandstone by the American geologists, who often designated formations by their lithological characters only, as Calciferous sandrock, Utica shale, Medina marl, Niagara limestone, etc. It frequently happens, however, that in the extension of strata into other regions their character changes, or they include beds for which the original lithological name would not be suitable. Sir Wm. Logan therefore considered it better to apply the term formation in all cases. The Potsdam formation skirts the borders of the Laurentian area in the counties

Hard sandstones.

Geographical distribution.

*Dr. Bell in Report of Geological Survey for 1876, p. 214.
†See Report of the Geological Survey for 1877.

of Frontenac, Leeds, Lanark and Carleton, and is well exposed in the townships to the north-east of Kingston and in many places between the Thousand islands and the Ottawa river. Its total thickness in this part of the province has not been ascertained, and it is variously estimated at from 300 to 700 feet. Fossils are not abundant in these rocks, and shells of the genus *Lingula* are perhaps the most characteristic. Large trilobites are occasionally met with, and at Perth certain remarkable tracks, supposed to have been made by these animals when the present hard rock was in the state of soft sand, have been found on the surface of one of the beds. Tracks left by creatures of a similar kind were found by the late Mr. Robert Abraham in 1847 in beds of the Potsdam sandstone near Beauharnois, in the province of Quebec. In the sandstones near Perth the late Dr. Wilson, nearly thirty years ago, found a number of long cylindrical casts like tree trunks from six inches to one foot in diamater. Last year attention was called to certain cylindrical bodies of larger size than the above which pass almost at right angles through the sandstone beds of this formation near the Rideau canal about eight miles from Kingston. The only economic materials furnished by the Potsdam formation consist of sandstones for building and glass-making. They are all too hard for grindstones or scythe-stones. The parliament buildings at Ottawa are constructed of Potsdam sandstone from the adjoining township.

Fossil tracks.

Economic materials

Lake Superior region.

In the lake Superior region the sandstones of Sault Ste. Marie, the peninsula between Goulais and Batchawana bays, Isle Parisienne, etc., seem to be of Potsdam age. They are mostly red, with green spots thickly sprinkled over the bed-planes, and interstratified with greyish layers. Unlike the Nipigon formation, they appear to be free from local disturbances and lie almost flat. Although they resemble some of the sandstones of Namainse in being red, they are believed to be newer and are probably unconformable to them.

THE SILURIAN SYSTEM.

Extent and character of the Silurian system.

This system was named by Sir Roderick Murchison after the Silures, a people who inhabited a part of ancient Britain in the border land between England and Wales. It is one of the most important systems in the geological scale, occurring in nearly all quarters of the globe, and is remarkable for the uniform character of its fauna in widely separated countries. It is almost everywhere rich in fossils, which consist principally of the remains of marine invertebrate animals and marine plants, although fishes and some land plants make their appearance in the upper part of the system. These rocks were formerly divided by some geologists into a Lower and an Upper Silurian series. The former is now often called the Ordivician, thus restricting the term Silurian to the upper division. For the present, however, we will retain the name Silurian for the whole system. The Silurian rocks appear to have been deposited during a generally quiet period of the earth's history. They embrace every variety of sediments, and occasionally include some igneous intrusions and beds of volcanic origin. They are divided into a number of formations, and the total volume of the system is very great in most regions. The thickness of each of the formations in Ontario will be given separately.

CALCIFEROUS FORMATION.

The name of these rocks is derived from their lime-bearing character. The formation is not important in Ontario, and is found principally between Brockville and Ottawa. It has a thickness of about 300 feet and consists for the most part of a bluish grey magnesian limestone, which has a gritty feel like sandstone, especially on weathered surfaces. The Ramsay lead vein near Carleton Place cuts this formation. Locality of the formation.

CHAZY FORMATION.

This formation derives its name from a town in Clinton county, in the state of New York. It is not an important formation in Ontario, and is found principally in the valley of the Ottawa below Pembroke, and between this river and the St. Lawrence below Prescott. Two outliers occur in the county of Renfrew, one on the Bonnechere and the other on the Madawaska river. It consists of greyish limestones, sandstones and shales, and has a thickness of about 150 feet. Some of the Chazy limestones are very suitable for building, and in certain localities the sandstones are also used for this purpose, but they are generally rather too thinly bedded. Geographical distribution.

BLACK RIVER AND BIRDS-EYE FORMATION.

The formations known under the above names in the state of New York are not regarded in Canada as differing sufficiently, either palæontologically or lithologically, to require separation. The Black River formation derives its name from a stream which enters the eastern extremity of lake Ontario in the state of New York, while the term Birds-eye has reference to the appearance of a fossil contained in the rocks bearing this name. The united formations have in Ontario a thickness of 150 to 200 feet, and consist of bluish and dark grey bituminous limestones with interstratified grey shales. It occurs on some of the islands in the north channel of lake Huron between the Manitoulin group and the north shore. Further east it skirts the southern edge of the Laurentian area from Penetanguishene to Kingston, and it is found in patches in the Ottawa valley above the city of the same name, and as a border surrounding the Trenton basin between the Ottawa and St. Lawrence further east. It is well developed around Kingston, and the building stone of the Limestone City is derived from it. Part of the stone used at Ottawa and Cornwall are quarried from this formation. The lithographic stones of the Marmora and Madoc region also belong to it. Origion of the names. Geographical distribution. Economics.

TRENTON FORMATION.

This important set of rocks is named from Trenton in the state of New York. On lake Huron it is found on Lacloche island and about Little Current in the northern part of Grand Manitoulin island. It occupies a broad belt between Georgian bay and lake Ontario, extending from Matchedash bay to Collingwood harbor on the former and from Newcastle to Amherst island on the latter. Lake Simcoe is situated entirely on this formation, and the whole of the peninsula of Prince Edward is underlaid by it. There is a Trenton outlier in the county of Carleton and it forms the uppermost rock in a geological basin, occupying the whole width of the country Geographical distribution.

between the St. Lawrence and Ottawa east of Ottawa city. The higher parts of the limestone cliffs at the capital belong to this formation. Judging from the results of borings which have been made in various localities, as well as from its general regularity and persistence, the formation is supposed to extend at a moderate but increasing depth south-westerly under the whole of the peninsula between lake Huron on the one side and lakes Erie and Ontario on the other. It has also been shown by borings in Ohio to underlie the newer rocks over a large part of that state. In Ontario it has probably a total thickness of about 600 feet and consists of fossiliferous bituminous limestones, usually dark grey in color, interstratified in some parts with shales which are also often bituminous. It affords excellent building stones in almost every part of its distribution, and it is important as a source of petroleum and natural gas. The oil and gas of the comparatively new field in north-western Ohio are derived from these rocks. The writer has shown that the Cincinnati anticlinal, along which these products have accumulated, continues northward and crosses the western part of the Ontario peninsula, following a line from near Little's point on lake Erie to near Kettle point on lake Huron.* By an inspection of the map it will be seen that Kingsville, where a valuable gas well has been recently struck, is situated on this line. Both the petroleum and gas of this formation have probably originated from the decomposition of the remains of marine vegetation, of which there is abundant evidence in these rocks. The most promising places for boring for either oil or gas would appear to be on the lines of the principal anticlinals, but only where the formation is well covered by impervious strata which have had the effect of confining these products for ages. Surface indications may be entirely wanting.

It underlies the western peninsula.

Petroleum and natural gas.

Cincinnati anticlinal.

Kingsville gas well.

Where to bore for gas.

UTICA FORMATION.

Although this formation is only about 100 feet thick in Ontario it is easily recognised, consisting everywhere of a black bituminous shale. The name is derived from the town of Utica, in the state of New York. It occurs on the northern points of Grand Manitoulin island and on the south side of Clapperton island. It is well seen to the west of Collingwood harbor, and runs thence south-eastward through the country, coming out on lake Ontario between Whitby and Newcastle. The Utica shales are sufficiently bituminous to burn with flame for a short time when thrown upon a hot fire. In October, 1859, works were erected a short distance west of Collingwood for distilling illuminating oil from these shales, which were found to yield from 3 to 4 per cent. of their weight of tarry oil at a cost of 14 cents per gallon. Owing to the discoveries of free petroleum the following year, this enterprise was abandoned. Illuminating oil is, however, still made at a profit from bituminous shale in Scotland.

Geographical distribution.

Shale oil.

* "The main axis of the anticlinal will intersect the north shore of lake Erie in the vicinity of Little's point in the county of Essex; then running about north-north-east through Essex, Bothwell and Lambton, it will reach the southern shore of lake Huron near Kettle point. Its general bearing from lake Erie to lake Huron is about north thirty degrees east, but it appears to curve gently to the south-east of a straight line and to pass under Petrolia." Transactions of the Royal Society of Canada for 1887, page 107.

HUDSON RIVER FORMATION.

This formation, named after the Hudson river in the state of New York, has a thickness in Ontario of about 700 feet and consists of bluish drab marls, clays and shales, interbedded with layers of limestone and sandstone. It is met with along the northern part of Manitoulin island and on the south-western side of Georgian bay, and it extends thence south-eastward through the country, widening as it goes, to lake Ontario, on which it occupies the shore from Port Credit to Pickering. An outlier of the formation, eighteen miles in length, occurs in the counties of Carleton and Russell a short distance south-east of the city of Ottawa. The Hudson river formation has furnished no economic minerals of importance in Ontario. The greater part of these rocks were at one time called the Lorraine shales, but that name is now abandoned.

Nature of the formation.

On lakes Huron and Ontario.

Outlier near Ottawa.

MEDINA FORMATION.

The Medina formation is named after Medina in the state of New York. It consists of red with some green marls and a fine grained light grey and sometimes reddish sandstone, called the grey band at the top. In the west the formation first appears near Colpoy's bay, on the south-west side of Georgian bay, and increases to the southward. It has acquired a thickness of about 200 feet in the eastern part of the county of Grey. Continuing thence southward it crosses the country, the thickness still increasing, till it strikes lake Ontario, where it amounts to about 600 feet. Its lower or eastern side comes out upon the lake near Port Credit, and the formation continues thence westward doubling around the head of the lake at Dundas, from which it runs along its southern shore eastward to the Niagara river and crosses into the state of New York. The sandstone at the top of the formation is an excellent building stone, and it is also used for grindstones and scythe stones. This band begins in the township of Nottawasaga, and is found all along the course of the formation to lake Ontario. Some beds of a brownish pink color occurring at the Forks of the Credit are highly esteemed as building stones.

Red marls.

Grey band.

On lake Ontario.

Forks of Credit.

CLINTON FORMATION.

This is named from Clinton county in New York state, and consists in Ontario of greenish and drab grey shales and thinly bedded siliceous and argillaceous limestones of similar colors, amounting to from 80 to 180 feet in thickness, together with a very ferruginous red band which, near Rochester, is called the "iron-ore bed," where it is said to have been used at one time as an ore of iron. The Clinton formation runs lengthwise through the centre of Manitoulin island, along the south-west side of Georgian bay, and thence southward to the head of lake Ontario, from which it strikes eastward along the base of "the mountain" and crosses the Niagara river. In the county of Grey the "iron-ore bed" is bright red and chalky or marly, but near lake Ontario it has become harder and more shaly, and contains a somewhat larger percentage of iron.

Nature of formation.

Distribution.

Iron-ore bed.

NIAGARA FORMATION.

This is one of the best marked of the fossiliferous formations of Ontario. It runs through all the Manitoulin group of islands, the Indian peninsula and thence to the Niagara peninsula, crosses the Niagara river and ends in Herkimer county in New York. It appears to attain its maximum thickness on Grand Manitoulin island, where the writer estimated it at 450 feet. At Owen Sound it is about 400 and at Hamilton 240 feet. It thus diminishes towards the south and east, while the underlying formations increase in these directions. Except along the Niagara river, where the lower 80 feet consist of bluish black shale, the formation is made up of dolomite or magnesian limestone. Northward of lake Ontario it becomes thickly bedded, of an open crystalline texture and a light grey color, but in the Niagara peninsula it is of a darker shade, closer texture and is more thinly bedded.

Thickness.

Nature of formation.

The Niagara formation is remarkable for the prominent escarpment which marks the lower or eastern boundary in all parts of its distribution. It is a conspicuous feature all along the sinuous course of the base of the formation to the south-west of Georgian bay, and forms the upper part of the Blue mountains in the townships of Collingwood and Osprey, which have an elevation, according to levels taken by the writer, of upwards of 1,200 feet over lake Huron, or about 1,800 feet above the sea. This is higher than the average altitude of the watershed between the great lakes and Hudson bay. From the Blue mountains the escarpment follows a general southerly course to the head of lake Ontario, and from thence it forms the crest of "the mountain" as far as Queenston. The gorge of the Niagara river, into which the falls pour their waters, cuts through the formation, the upper or limestone part amounting to 164 feet in thickness, and the above mentioned shale at the bottom to 80 feet. The Niagara limestone everywhere in western Ontario affords an excellent building stone, and it also burns to good lime.

The Niagara escarpment.

Blue mountains.

Niagara Falls.

At the head of lake Temiscaming, which is situated at the great bend of the Ottawa river, there is a large outlier of this formation consisting of from 300 to 500 feet of grey limestones, with arenaceous beds and coarse or boulder conglomerates at the base.

On lake Temiscaming.

In the northern part of the province, west of James bay, we meet with almost horizontal grey and yellowish-grey limestones, containing fossils, which, according to the late Mr. E. Billings, the celebrated palæontologist, belong to the Niagara formation. These strata occur along the Albany river above its junction with the Kenogami, and also along the latter stream as far up as the first portage. The limestones are overlaid by a considerable thickness of chocolate-colored marls with greenish layers and patches, but without observed fossils.

Near James bay.

GUELPH FORMATION.

This formation, which occurs only in Ontario, was named at the suggestion of the writer after the town of Guelph, which is situated upon it. Its greatest thickness, about 160 feet, is attained in the central part of the western peninsula, from which it diminishes both south-eastward and north-westward, terminating about the Niagara river in the one direction and on the south side of the Manitoulin island in the other. Throughout the greater part of its distribution it consists of a light buff or cream-colored dolomite of a finely crystalline or granular texture, resembling sandstone, but in the Niagara peninsula it becomes dark grey and bituminous, and more distinctly

Thickness and distribution.

crystalline. It is well defined as a formation by a considerable number of characteristic fossils. The Guelph dolomites form beautiful building stones, and they have been largely used for this purpose in Galt, Guelph, Elora and Fergus. They also burn to lime of excellent quality. Economics.

ONONDAGA (SALT) FORMATION.

This formation is named after Onondaga in the state of New York, and is celebrated for its salt-bearing character. It consists principally of yellowish and drab colored dolomites and greenish and drab shales, with some reddish layers, especially near the base of the formation. It occurs along the east shore of lake Huron from Goderich to the mouth of the Saugeen river, from which it turns east and south, rounding the northern end of a wide synclinal between Southampton and the head of Owen Sound, and running thence south-easterly to the Grand river, from which it takes an easterly course to the Niagara. The numerous borings which have been made through the formation in search of salt in the country to the east of lake Huron prove it to have a thickness of 775 feet at Goderich and 508 feet at Kincardine, but this has diminished to about 300 feet where it crosses the Niagara river above the falls. The beds of rock-salt which furnish the brine of the wells at Kincardine, Wingham, Blyth, Clinton, Goderich, Exeter, Seaforth, etc., occur towards the base of the formation and are only reached by deep boring. The bore-holes for some of these salt wells have also passed through deposits of gypsum. Beds of this mineral occur likewise along the Grand river from a short distance above Paris to near Cayuga. Most of it is of a grey color, useful as a mineral manure, but in some places it is white enough to calcine for stucco and alabastine. One of these localities is the Merritt mine, where there is a bed of white gypsum four to six or seven feet in thickness. Further particulars of these deposits, contained in the evidence collected by the Commission, are published in another part of this report. Nature of formation. Distribution. Thickness. Salt. Gypsum.

In this formation on the east side of the Saugeen river, just above Walkerton, the writer in 1861 discovered lithographic stone of excellent quality, but breaking transversely into pieces of too small size to be of much value. The band forms the top of the bank of the river, and the beds associated with it burn into a remarkably white lime. Lithographic stone.

On Moose river, banks of gypsum occur from ten to twenty feet high, especially on the north-west side below the junction of the Missinaibi, for a space of about seven miles, or from thirty-one up to thirty-eight miles above Moose Factory. About ten feet of the lower part of the deposit consist of solid gypsum of a light bluish grey color, but the upper portions are mixed with marl. In some sections of these banks a comparatively small proportion of the gypsum, but still large commercially speaking, is nearly white, and from this circumstance they have received the name of "the white banks." The geological age of these deposits cannot be far from the Onondaga formation, and it would not be surprising if salt should also be found in the rocks with which they are associated. Gypsum on Moose river.

LOWER HELDERBURG FORMATION.

Waterlime division.

A portion of the Waterlime division of the Lower Helderburg formation of the state of New York reaches the township of Bertie on the Ontario side of the Niagara river, but as it is unimportant and closely connected with the Onondaga formation, it requires no further description in the present short sketch.

THE DEVONIAN SYSTEM.

Old and New red sandstones.

This system, which derives its name from Devonshire in England, was at one time called the Old Red Sandstone to distinguish it from the Permian, which was known as the New Red Sandstone. Some geologists have advocated changing the name in America to "Erian," but there seems to be no sufficient reason for this, and as Sir William Logan followed the method of calling the systems by their British names and the formations by those adopted in the United States, we prefer to adhere to an uniform plan of nomenclature, and to continue to recognise the well established name of this system. The Devonian rocks are important in various countries, from holding deposits of petroleum, salt, gypsum and iron ore, and they are also of interest geologically from the fact that it is among them that the remains of fishes and land plants first became abundant. Red sandstones form a prominent feature in the Devonshire rocks of the eastern part of the Gaspé peninsula in Quebec, but they are absent from them in the province of Ontario. This system occupies a considerable area in the western peninsula and again in the northern part of the province, and in both of these regions the Corniferous formation constitutes its most prominent member.

Economics.

Red sandstones.

ORISKANY FORMATION.

Sandstone.

This is the lowest in the Devonian system, and is represented in Ontario by only about twenty-five feet of coarse grey and brownish sandstone. It is exposed in various places along the base of the next higher formation between the township of Windham and the Niagara river. It has been used as millstones for the preparation of oatmeal, and also for building purposes.

CORNIFEROUS FORMATION.

Distribution.

The Corniferous formation is so called from the nodules of hornstone which it frequently encloses. Its base or lower border runs north-eastward from near Goderich to the township of Greenock, where it turns around the north end of the wide synclinal already referred to, from whence it takes a southward course to the township of Burford, and then strikes eastward to the township of Bertie. The shore of lake Erie from the outlet of the Niagara river to Port Rowan lies upon this formation. Its junction with the overlying Hamilton formation is covered with superficial deposits, but it is supposed to run northward from near Port Rowan, keeping at a distance of twenty to twenty-five miles from the line above described as marking the base of the formation.

Nature of formation.

Thickness.

In western Ontario it consists mostly of grey limestone, containing great numbers of fossil corals, some of which form masses of considerable size. It makes a fair building stone, and is also burnt for lime. Borings for wells in south-western Ontario have given the following thicknesses for limestone supposed to represent the Corniferous in each case: Port Lambton, 320 feet; Petrolia, 248 and 378; one mile south-west of Belle river, 209; Leamington, 310. In such borings, however, it is difficult to draw a line between the limestones of this formation and those of the underlying Onondaga.

The petroleum of the Enniskillen field is drawn directly from the Corniferous limestone, but it has not been proved that it originated in these rocks. There are reasons for supposing it to be quite as likely that the oil comes up from the underlying Trenton formation.* Petroleum.

In the region south-west of James bay the Corniferous formation occupies an area greater than all the western peninsula of Ontario. A large part of this, lying between the Albany river and the basin of the Moose river, comes within the northern part of the province. It consists mostly of porous and cavernous drab grey and yellowish grey fossiliferous limestones resting directly upon the Archæan rocks to the southward, the line of junction cutting the Missinaibi river just below Hell-gate, the Mattagami just below the Long Portage, and the Abittibi just below "The Otters" portage. Many of the Corniferous fossils of this district belong to species which differ from those of the formation in regions to the south of the height-of-land, tending to show that there was here a separate basin in these early times as well as now. At the foot of Grand rapid on the Mattagami river the writer, in 1875, discovered a large deposit of rich clay-ironstone in these rocks.† The materials of the Drift, for a considerable distance to the southward of the Corniferous formation in this region, contain fragments of this ore, indicating that it exists, and probably in the same horizon, among these rocks in many other places besides the above mentioned locality on the Mattagami. James bay region. Clay ironstone.

HAMILTON FORMATION.

This is not called after the city of Hamilton, but a village of the same name in the state of New York. It consists of bluish and drab grey clays or marls, called "soapstone" by the well-borers, with some greyish limestones, and occasionally an arenaceous band. The total thickness of the formation in Ontario is estimated to be about three hundred feet, of which the lower 170 or 185 feet are found below the Drift clay and above the Corniferous limestone in the oil territory of Enniskillen. This impervious rock has served to prevent the upward escape and loss of the petroleum and gas of this region in past ages. Nature of formation

CHEMUNG AND PORTAGE FORMATION.

This is represented in Ontario by a few feet of black bituminous shales in the southern part of the county of Huron and the northern part of Lambton. A narrow border of the formation may also exist beneath the Drift on the north shore of lake Erie, between Rondeau and Port Talbot. In the states of New York, Pennsylvania and Michigan these rocks are, however, extensively developed, and constitute an important formation. Black shales.

*"The Petroleum Field of Ontario," by Dr. R. Bell in the Transactions of the Royal Society of Canada for 1887, page 109.

†Report of the Geological Survey for 1875-76, page 321.

THE POST TERTIARY SYSTEM.

The Post Tertiary system. The rocks which have been described in the foregoing pages comprise all the ancient or fundamental formations represented in Ontario, the remainder of the geological scale which is so largely developed in various other parts of the world being entirely wanting till we come to the Post Tertiary system, which includes our superficial deposits such as boulder-clay, stratified clay, sand, gravel, etc. The oldest of these is called the Drift.

THE DRIFT.

Evidences of glacial action. In a previous part of this section a description was given of the extensive glaciation which took place in the Archæan regions of Ontario during the Drift period, so that it will be unnecessary to dwell further on that part of the subject. The glacial phenomena are also very noticeable throughout the Palæozoic districts, so that everywhere in the province the surfaces of the solid rocks bear the ancient ice-marks in the shape of flutings or furrows and grooves or striæ. With the exception of the high lands near the east coast of Labrador, no part of the Dominion on this side of the Rocky mountains, as far as known, appears to have escaped the action of glaciers in the Drift period. The rocks in the Archæan districts are everywhere ground down and rounded, the evidence of the glacial action being usually as plain on the tops and sides of the hills as in the valleys. In the Palæozoic regions, where the strata lie almost horizontally, the wearing down of the rocks has taken place principally along the planes of bedding. Where the dip happens to be in about the same direction as that which was taken by this great denuding force, the excavations naturally deepened until a point was reached where the weight and solidity of the opposing rocks became sufficient to resist the ice-mass, and in this way escarpments have been formed. Escarpments, how formed. All the great lakes of the St. Lawrence, except lake Superior, lie in basins of erosion which have been hollowed out in the same manner. The basin of lake Superior, although its origin was of volcanic nature, has been much enlarged by glacial denudation. It has been shown that the lakes of our Archæan regions are all of glacial origin, and that most of them lie in rock-basins excavated during the Drift period. A few of them owe their existence to moraines or dams of glacial debris, which hold up their waters.

Origin of deep bays. The fracturing of the sedimentary rocks along anticlinal lines has greatly aided glacial erosion, and in this way long bays have been formed in the geographical outlines of the formations, such as those on Manitoulin island, the Indian peninsula and thence to lake Ontario, and all along the base of the Black River formation from Matchedash bay to Kingston.

A marked difference is observable in the effects of glacial action on the opposite sides of the Archæan nucleus on which the Palæozoic strata rest. Valleys or water channels have been formed wherever the ancient glaciers plunged downward off the Archæan highlands upon the opposing edges of the newer rocks, as all along the southern boundary of the Laurentian and Huronian rocks of the province. But no such action took place when the glacial mass was forced up the gentle slope of the Palæozoic beds of the basin of Hudson bay and thence upon the Archæan plateau to the south of it.

Here we find no physical features to mark the line of contact between the two kinds of rock which differ so much from one another. On the east side of Hudson bay deep channels and valleys with high escarpments facing inland have been formed by the descent of the old west-moving glaciers against the up-turned edges of the Cambrian rocks along that coast, while on the opposite side of the bay they moved off the Devonian and Silurian rocks without leaving any impression on the geographical features of that region.

Effects of the ancient glaciers.

In the metamorphic regions in the northern parts of Ontario, the rounded glaciated surfaces of the tops and sides of the hills have been left almost or quite bare in many parts, but in most districts and especially in the Palæozoic areas of the province the smoothed and grooved or striated rock-surfaces are covered by a thick deposit of stiff clay mixed with sand, gravel, stones and boulders. This is known as drift, boulder-clay, hard-pan, etc. In Scotland it is called till, and this convenient name is now being adopted in America and elsewhere. On the higher grounds north of lakes Superior and Huron there is usually but little clay, the drift consisting of loose boulders, stones, gravel and sand.

Drift in the metamorphic and palæozoic regions.

The transportation of the boulders in the till, as well as those lying on bare surfaces, has been simultaneous with the planing and grooving of the rocks, and due to the same cause. An erroneous impression which is very prevalent attributes both these phenomena to icebergs. Although the latter may have brought some boulders and dropped them among the Post Tertiary clays and sands, they appear to have had little influence on the formation of the underlying till, and they have had nothing to do with the wearing down and grooving of the solid rocks. The ice-grooves are locally nearly parallel, except in cases where different sets cross one another. In pursuing their course they will go up one side of a rounded ridge or knoll of rock and down the other, or they may curve around it and even pass under overhanging rocks, grooving both the wall and roof in a manner quite impossible to have been produced by a floating iceberg. The glacial phenomena of the Drift period in these latitudes correspond in every way with what may be observed on a small scale in connection with modern glaciers, and there can be no doubt that they have been due to land ice. These phenomena occurred at this period in the north temperate zone all around the globe, and the gigantic scale on which they operated constitutes one of the most extraordinary phases of the earth's history. The prevalence of ice was so general at this time that it is also known as the Glacial Period.

Glacial phenomena.

The general direction of the glacial movement over Ontario, as shown by the striæ, was southward, but it varied greatly to the east and west of south in different regions. North of lakes Huron and Superior, and from the latter westward to Lake-of-the-Woods, it was generally south-westward, but in some instances it varied greatly from this on account of local causes. In the western peninsula it was south-eastward, but around lake Ontario south-westward; in the lower Ottawa valley south-eastward, but north of it the direction was south-south-westward. In the Eastern Townships it was south-eastward, while around Montreal the course was south-westward.

Direction of the glacial movement.

The striæ following the above courses may not have been all produced at the same time and by a continuous glacier. The ice-sheet would probably move in different courses in different parts, according to the general slope of the surface on which it rested, or according as it accumulated in one part and the resistance became relieved in another. When the maximum had passed, the more its mass diminished the more it would be influenced by the local form of the land. Finally, when it became divided into separate glaciers, these would follow the valleys or would be guided by their confining ridges. Hence in the bottoms of many valleys we find the striæ parallel to their general trend. There is reason to believe that the relative levels of some parts of this continent have changed considerably since the Drift period, and this circumstance must be taken into consideration in connection with the formation and the movements of the ice-sheets of glacial times.

Local causes influencing the course of glaciers.

The local or final glaciers of the period sometimes ploughed their way into the mass of till which had been left by the more general one. They also left behind them lateral ridges or moraines of boulders and earth. Some fine examples of these are to be seen on either side of the southern part of Long lake, north of lake Superior, and along the upper parts of the valley of Steel river in the same region. In some cases the ancient glaciers also left terminal moraines, and these by damming up the waters have formed some of our smaller lakes in the north country.

Lateral and terminal moraines.

At any given locality the greater part of the materials of the drift usually consist of the debris of the rocks immediately underlying it, but it generally also contains a large amount of transported material, the percentage diminishing about in proportion to the distance from which it has been carried, the harder rocks surviving the wearing action the longest, and thus travelling the furthest.

Material of the drift.

On the generally lower levels of the province, and in local depressions elsewhere, we find stratified clay, sand and gravel resting upon the till. These sands and gravels are usually above the clay. It is supposed that the cause of this was a submergence of the land after the Glacial period, during which the clays were deposited, and as the land rose again the sands were spread over them, and that both deposits were worn into terraces during stationary intervals while the general elevation was going on.

Clay, sand and gravel strata.

In the eastern and northern parts of the province some of the clays and sands contain sea shells and other fossils, indicating a marine origin. In the valley of the St. Lawrence these are found as far west as Brockville, and along the Ottawa they extend about as far up as the junction of the Bonnechère river in clays and sands which constitute continuations of extensive deposits of the same character in the province of Quebec. But no marine fossils have as yet been found in any of the Post Tertiary deposits in the province west of these points and south of the watershed of Hudson bay. The writer has, however, discovered a variety of marine shells on the Albany, Kenogami, Missinaibi and Mattagami rivers up to heights of about 300 feet above the sea level and more than one hundred miles inland.

Marine fossils in districts of drift.

West of the points above mentioned, south of the height-of-land, the marine deposits are replaced by others which appear to be in part, at least, of fresh water origin. One of the most important of these is an extensive blue clay deposit which we have called the Erie clay, and which has as yet yielded no organic remains of any kind. It burns to white bricks, while the marine clays to the east burn red. The Erie clay is often very calcareous, and is seldom or never entirely free from pebbles and stones, more or less thickly disseminated through it. Indeed it often seems to merge into the underlying boulder clay. It covers the whole of the south-western part of the western peninsula, and is locally developed in many other parts of the province as far east as the line of railway from Brockville to Ottawa. Its greatest known depth is about 200 feet, but it is found at differences of levels amounting to 500 feet. When seen in fresh section it presents lines of stratification, and often a transversely jointed structure. In some localities its upper parts have been unevenly denuded before the deposition of the next higher formation, which consists of brownish clay yielding red bricks. This unconformable formation is well developed in the valley of the Saugeen river, and hence it has received the name of the Saugeen clay. Its thickness appears to be less than that of the Erie clay, but it is found in broken areas in all parts of the province except the most easterly and northerly. When seen in fresh section it is usually found to be very distinctly stratified in thin layers, sometimes with partings of fine sand between them. Beds of sand and gravel are occasionally found between the Erie and Saugeen clays, and these are of importance as affording good wells of water. Fresh-water shells have been detected in a few instances in the Saugeen clay. The Erie and Saugeen clays.

The sand deposits overlying the Saugeen clay in the southern parts of Ontario are too irregular and varied in character to admit of classification for the present. But in the district of Algoma and between the great lakes and the Ottawa river a yellowish sand, to which the name of the Algoma sand has been given, is extensively distributed in the more level areas, while on the higher grounds are found considerable accumulations of gravel, stones and boulders, which have been already referred to. Deposits of clay resting on sand with clay again beneath are found over large areas in the extensive and comparatively level tracts beyond the height-of-land. These regions have been explored and reported upon for the provincial government by Mr. E. B. Borron, who has paid much attention to their surface geology. Algoma sand.

In the western peninsula there is a remarkable and very extensive accumulation of gravel above or west of the Niagara escarpment, which extends from near Owen Sound to Brantford. It has been called the Artemesia gravel, after the township of that name, and consists principally of the debris of the Niagara and Guelph formations, with some pebbles and boulders of Laurentian origin. The gravel, which has a considerable depth, is well rounded, often washed clean of finer material, and is extensively used for road metal. Artemesia gravel.

From an economic point of view the superficial deposits are important in relation to water supply, the nature of the soils which they afford, etc., and many of the clays have a direct value for the manufacture of bricks and drain tiles. The shell-marls and peat among the recent deposits also belong to this part of our subject. Lignite, associated with clay and sand, is found on the Economics of the Drift.

Goulais river, and indications of it have also been met with on Rainy river and the southern part of Lake-of-the-Woods. North of the height-of-land the writer has found beds of this substance associated with the till and the overlying deposits in several places on the Missinaibi river, and also on the Kenogami.

THE COPPER AND NICKEL MINES OF THE SUDBURY DISTRICT.

Supplementary notes.

Since the foregoing chapter was written, considerable progress has been made in the development of the mineral wealth of the Sudbury district. Besides the Stobie, Copper-cliff and Evans mines, belonging to the Canadian Copper Company, which have been steadily worked and have yielded a large amount of copper and nickel ore, two other mines have been in operation and some new localities have been discovered. The writer has had opportunities for further study of the geological and lithological relations of these deposits, and the following notes are added to bring the subject up to date. It will be seen from these that the copper and nickel ore deposits of the district resemble one another closely, and that they all appear to occur under similar geological and lithological conditions.

The Dominion Mineral Co's. property.

The deposit which had been discovered on lot 4, range 2, of Blezard, about one mile north of the Stobie mine, has been acquired by a new organisation called the Dominion Mineral Company, and is being vigorously worked. Three shafts are being sunk, each of which had reached a depth of about 40 feet in the middle of October. The ore consists of a body of mixed chalcopyrite and nickeliferous pyrrhotite mingled with more or less rock matter, giving the whole the appearance of a conglomerate. The general strike of the country rocks is here, as elsewhere in the vicinity, about north-east. The ore-bearing belt, which is associated with a dark quartz-diorite, is about 100 feet wide and dips north-west at an angle of 65 degrees. It is overlaid by a massive bed of ash-colored greywacké, the weathered surfaces of which present raised reticulating lines. Immediately to the north-west of the shafts there is a dyke from 30 to 50 feet wide, of dark-brownish grey crystalline diabase, weathering at the surface into rounded boulder-like masses, which scale off concentrically. At the place just indicated, the dyke runs south 35° west (mag.), but a short distance to the south-westward, what looks like its continuation, runs south 70° west and appears to be thrown a short distance northward by a dislocation.

Parallel ranges.

To the south-west of the Dominion mine similar ore has been found in the southern part of lot 5, in the 2nd range (Russel's), and also in the northern part of lot 6, in the 1st range (Stobie's) of Blezard. A large diabase dyke runs near the latter, and both discoveries are near the north side of a quartz-syenite ridge, which runs in a north-easterly course from the township of Snider, and appears to terminate before reaching the Dominion mine. The copper deposits of this mine, Russel location, Stobie location, Murray mine, McConnell mine, lot 10 in the 1st range of Snider, and lot 1 in the 1st range of Creighton would, therefore, all appear to be in the same run, on the north-west side of the syenite and gneiss belt, while the Stobie mine, the

Frood, lot (number 7 in the 6th range of McKim), the Copper Cliff mine and an outcrop of copper ore on lot 1, range 2 Snider, and another on lot 7, in the 6th range of Waters, would occupy a corresponding horizon on the opposite or south-east side of the ridge.

The Murray mine, situated on the northern part of lot 11 in the 5th range of McKim, and on the main line of the Canadian Pacific railway, 3½ miles north-west of Sudbury Junction, was prospected under a bond, by the Messrs H. H. Vivian and Company, of Swansea and London, and purchased by this firm on the 1st of October. At this locality the general strike is also north-easterly, and the ore body, which conforms with the stratification, is traceable for about quarter of a mile. A short distance south-west of the railway track, which crosses the north-eastern part of the deposit, it has a width of upwards of 100 feet. Here, as elsewhere, the ore is a mixture of chalcopyrite and nickeliferous pyrrhotite with incorporated masses of rock of all sizes, the deposit being, in fact, in all respects like those of the Stobie, Dominion and other mines. It is flanked on the north-west side by a very crystalline massive grey diorite, and on the south-east side by a dark greenish mottled variety of this rock, followed by alternate belts of more or less fissile amphibolite or hornblende rock and reddish grey quartz-syenite. Ten of these alternations occur in a breadth of 150 yards. A cutting on the railway at about this distance south-east of the mine, shows what may be either a conglomorate or concretionary mixture of these two rocks. The main ridge of quartz-syenite lies about half a mile to the eastward. A large dyke of crystalline diabase, with a west-north-western course and weathering into rounded masses may be seen on the railway just north-west of the mine, and again to the eastward a short distance north of the track. This and a parallel diabase dyke are seen in other places along and near the track between the Murray mine and Sudbury, and similar dykes occur on Ramsay lake, which may be continuations of these.

The Murray mine.

Diabase dykes.

Late in the summer a discovery of mixed chalcopyrite and pyrrhotite, of which large and fine specimens were shown to the writer, was made at a spot situated about three-quarters of a mile north-east of the bay of Wahnapitæ lake lying on the north side of the point and ridge between the east bay and the main lake. The deposit, which is said to be large and promising, is flanked by diorite on the south-east side. Those who discovered this ore-mass state that the surface indications consisted of nothing more than a black discoloration of the rock, and that it might have been easily passed over unnoticed. Another discovery of mixed chalcopyrite and nickeliferous pyrrhotite was made at the north-east end of Waddel's lake, which is the first of the small lakes on the canoe-route west from the western extremity of Wahnapitæ lake. The ore occurs in diorite, which comes in contract with quartz-syenite about half a mile to the south. Copper and iron pyrites, said to be nickeliferous, have been discovered on lot 3 in the 4th range of Levack. No work had been done at this locality at the time of the writer's visit, but the surface is covered by oxide of iron in the same way as at some of the other copper and nickel ore deposits of the district. The rock on the east side of the deposit is gneiss and on the west diorite. Similar ore is reported

Discoveries in the region of Wahnapitæ lake.

to have been found on lot 6 in the 2nd range of the same township, but the locality was not visited.

The Vermillion and Krean locations.

The copper deposit opened by the Vermillion Mining Co. on lot 4 in the 6th range of Denison, and that of the Krean mine, about a mile to the north of it, are both associated with brecciated diorite rocks, as well as the Copper Cliff and the Stobie, while gneiss or quartz-syenite occurs at a short distance to one side in every case. A deposit of copper ore, similar to these, occurs in the south-west corner of Snider (lot 10 in the 1st range), which is said to be associated with diorite and flanked on the south by syenite. At this locality there is also said to be a large dyke of crystalline diabase, weathering into the characteristic rounded boulder-like masses.

Occurrence of the nickeliferous copper ores in diorite.

It will be seen from the foregoing that all the deposits of nickeliferous copper ore of the district which have been examined, occur in diorite rocks, and further that in most cases the diorite is brecciated or holds angular and also rounded fragments of all sizes of rocks of various kinds, the prevailing varieties being other kinds of diorite, quartz-syenites, crystalline schists, greywackés and quartzites. The general geological position of these ores is therefore in diorite, and more especially brecciated diorite, with either gneiss or quartz-syenite near one side. The ore masses further resemble one another in having an appproximate lense-shaped outline, parallel to the general strike of the country rocks, although they may not always be strictly conformable with them in dip, but may take a different anlge to the horizon, as if they had been connected with longitudinal fissures and were of the nature of great brecciated veins or "stockwerks."

Age of the diabase dykes.

The occurence of dykes of crystalline diabase near several of the deposits of copper and nickel ore has been referred to. Some of these dykes run west-north-west, others south-west, and one at the outlet of Ramsay lake runs about west, or towards the Copper-cliff mine. These dykes cut through all the stratified Huronian rocks of the district, and also the quartz-syenites, whether they occur as narrow bands or large areas. They are, therefore, newer than any of these rocks and they are found on microscopic examination to be apparently identical with the diabase overflows of the Animikie formation of lake Superior. Their association with the ore deposits of this district suggests some connection with them ; and it may be found on fuller investigation that where they cut the ore-bearing belts they have had something to do with their enrichment at these places. If this should prove to be the case, it will be important to trace these dykes, as well as the stratigraphical horizons along which the metals may have been originally deposited in a diffused form, as new discoveries may be looked for at these inter-sections.

The smelting works.

The smelting furnace which had been erected at the Copper-cliff mine, towards the close of 1888, has been steadily in blast and has reduced an average of about 125 tons of roasted ore per day, the amount sometimes running up to over 150 tons. A second smelter, in every respect like the first, which was erected during the past summer (1889), commenced work on the 4th of September, and has now been running with equal success for two months. The Dominion Mineral company and the Messrs. H. H. Vivian and Company are erecting similar furnaces at their mines.

THE GOLD DISCOVERY ON LAKE WAHNAPITÆ.

The gold mining location on the south side of lake Wahnapitæ was again visited in the month of October of the present year. Some work had been done and the true nature of the auriferous rock may now be studied to more advantage than when the locality was visited last year, before the ground had been broken. An excavation, measuring about 20 feet in length by 8 feet in depth and the same in width had been made. This opening shows that the gold-bearing portion of the ridge of felsitic quartzite follows a belt of quarztite boulder-conglomorate, which runs south-westerly. Some of the individual masses are sub-angular, but most of them are rounded and they vary from a few inches up to 10 feet in their greatest diameters, which are parallel to the walls. A few rounded greenish, somewhat schistose masses are also included, and all are packed closely together. The interstices are filled with a rather coarse glossy, greenish to yellowish grey hydro-mica or talcoid schist, which, on weathered surfaces, is seen to be full of pebbles of bluish quartz and white quartzite, from the size of coarse gravel down to that of pease or smaller. The quartzite boulders vary in texture from granular to compact or cherty. In color they present shades of light, dark and reddish grey; also of greenish or olive grey. The last named contain bunches of crystals of mispickel. In the width of the excavation, there are four or five veins of white quartz from two to three inches in thickness, or aggregating about a foot. These show specks or small nuggets of free gold and Mr. Richardson, the manager of the mine, informed the writer that he had detected visible gold also in the schistose filling, as well as in the quartzite itself. A band of fine-grained dark colored diorite runs parallel to the quartzite ridge at no great distance on either side of it. That on the north-west side appears as if it had flowed upon an uneven surface of "lumpy" quartzite. An attempt to extract the gold from the quartz in this locality, by means of a small arastra, had been commenced at the time of my visit. The quartz was first calcined in a wood fire, after which it was easily ground under the flat surfaces of two large stones attached to a beam drawn round by a horse. These stones, worked upon a smooth pavement of smaller ones, surrounded by a circular wall, which held in a few inches of water with some pounds of quicksilver in the bottom.

Nature of the auriferous rocks on lake Wahnapitæ.

www.ingramcontent.com/pod-product-compliance
Lightning Source LLC
Chambersburg PA
CBHW022014190326
41519CB00010B/1518

* 9 7 8 3 7 4 4 7 1 7 7 5 5 *